Labor and the Environmental Movement

Urban and Industrial Environments

Series editor: Robert Gottlieb, Henry R. Luce Professor of Urban and Environmental Policy, Occidental College

Labor and the Environmental Movement

The Quest for Common Ground

Brian K. Obach

The MIT Press
Cambridge, Massachusetts
London, England

This book was set in Sabon by SNP Best-set Typesetter Ltd., Hong Kong

Printed on recycled paper and bound in the United States of America.

Library of Congress Cataloging-in-Publication Data

Obach, Brian K. (Brian Keith)
 Labor and the environmental movement: the quest for common ground / Brian K. Obach.
    p. cm.—(Urban and industrial environments)
 Includes bibliographical references and index.
 ISBN 0-262-15109-X (hc.: alk. paper)—ISBN 0-262-65066-5 (pbk.: alk. paper)
  1. Labor unions—United States. 2. Environmentalism—United States. 3. Coalition (Social sciences). 4. Interorganizational relations. 5. Social movements. 6. Pressure groups. 7. Organizational sociology. I. Title. II. Series.
HD6508.O2   2004
331.88′0973—dc22

                                                        2003059359

10   9   8   7   6   5   4   3   2   1

For İlgü

Contents

Preface

In 1990 I participated in some of the Redwood Summer demonstrations in northern California. These protests were organized in response to the increasingly rapid liquidation of the few remaining old-growth forests in the United States. I was aware of the tensions that existed between environmentalists and some members of the logging community, but I was comfortable participating in the demonstrations in that the clearly defined targets were the corporate owners of the logging companies. Nonetheless, I designed my banner with the plight of the workers in mind. It read: "I support workers AND the environment." But at some of these demonstrations I faced counterprotesters, workers from the local mills, men and women concerned about their jobs and resentful of the activists bussed in from San Francisco and other foreign lands.

This was deeply disturbing to me. For years I had been active in several social movements, from the struggle against apartheid to the war in Central America. I was and still am an active union member, and I had done a fair amount of labor support work as well. Through it all I never found myself on the opposite side of the picket line, the target of protest by the working-class people with whom I identified. The counter-demonstrators seemed disturbed by my sign as well, their anger shifting to confusion. Other environmentalists were already reaching out to the workers and the local logging communities. Some insightful activists had recognized the strategic and moral failure of not having done so sooner. Timber industry workers were not the enemy; they were important allies and fellow victims of the profit-hungry corporations that exploited workers with little regard for the long-term sustainability of their jobs or the environment. Short-term profits were the order of the day and everyone, including the workers, stood to lose.

Although some important relationships were built between workers and environmentalists through the struggle over preservation of old-growth forests, the experience of being the target of worker protest continues to haunt me. It was this experience that motivated me to dedicate myself to understanding the profoundly important relationship between unions and the environmental movement. In my assessment of the contemporary political moment, the quality of this relationship and the invigoration of both the labor and environmental movements will determine whether we as a society will create a just and sustainable economy or further the hyperexploitation of workers and continue down the path of ecological ruin. It is my greatest hope that in some small way this book will help to bring about the former.

There are many people who made this work possible. I offer a great deal of gratitude to the activists and movement leaders who took time away from their important work to speak with me. Several people were particularly helpful in this regard, including Jane Perkins, Bill Towne, Ken Germanson, Ned McCann, Monica Castellanos, and Bill Norbert. Several people offered helpful feedback on earlier drafts of this work including Gay Seidman, Fred Buttel, Gerald Marwell, Allen Hunter, Robert Gottlieb, Fred Rose, and Brad Barham. A special note of thanks to Pamela Oliver, who provided support and guidance throughout the course of this project and during my entire stay at the University of Wisconsin. I also appreciate the support of Annee Roschelle, Peter Kaufman, Hal Jacobs, Irwin Sperber, Susan Lehrer and my other friends and colleagues at SUNY New Paltz. I offer additional thanks to Clay Morgan, Deborah Cantor-Adams, and the other folks at the MIT Press.

I want to thank my mother, Linda Horowitz, for raising me right and providing constant encouragement. Thanks to my good friend Stuart Eimer for helping to discern what is real and important throughout this process. This book is dedicated to İlgü Özler for her unconditional support, love, and therapy, without which I never would have found the strength to finish.

The research for this project was funded in part by the National Science Foundation and the University of Wisconsin Department of Sociology Small Grants Committee.

Labor and the Environmental Movement

1

Introduction

Sandy Fonda is the chair of the Rainbow Alliance for a Clean Environment. Bill Towne is a district manager for UNITE, the Union of Needletrades, Industrial and Textile Employees. They both live and work in Fulton County, New York. Fulton County is a rural area of upstate New York about an hour northwest of Albany. Historically the economy of the region has been based on the garment industry. Gloversville, one of several small towns to dot the countryside, was the site of several tanneries that produced leather and leather products (including gloves, the town's namesake). Several tanneries still operate in the town, the pollution from which led to many years of conflict between Sandy Fonda and Bill Towne.

Problems began in the early 1980s when, according to Towne, "the world finally found Fulton County" (personal communication, May 10, 1999). Environmental problems caused by the tanning industry had been largely ignored for years. But by the 1980s a group of activists who had been working on nuclear issues turned their attention to local environmental problems. Led by Fonda, this small community environmental group engaged in grassroots actions targeting specific tanneries for their environmental violations. Rainbow Alliance members also pressured the federal Environmental Protection Agency (EPA) and the New York State Department of Environmental Conservation (DEC) to monitor the tanning industry in the area more closely and to investigate the industry's regulatory violations, which were numerous.

The first major conflict involving both Fonda and Towne arose when the DEC ordered the county to close and cap two local landfills, which were found to be contaminated with toxic waste from the tanning industry. The costs associated with this project were estimated to be in the tens of millions of dollars. The tanneries were to be required to pay for the capping and the construction of a new landfill, a cost they claimed they could not afford. Employers appealed to their workers for support in fighting the new requirements citing the threat of job loss if the tanneries were forced to bear this cost. According to Towne, the message from the industry was clear: "What was driven home to the union members was, 'They are going to shut us down! We are going to go bankrupt! We are going to close!' " (personal communication, May 10, 1999). These threats effectively turned the union against the Rainbow Alliance and the environmental measures they advocated. There were public confrontations between union members and environmentalists and between Towne and Fonda personally. Towne recalls, "It was truly white heat in those years. We were going to battle. . . . It took a number of years for that to settle. At the time, I'll speak for the union and the industry, we were dealing on a day-to-day basis and were not thinking long term at all. A lot of the members wanted things cleaned up, but we didn't want what we figured would be tremendous job loss" (personal communication, May 10, 1999).

While the landfill issue was being disputed, environmental regulators also determined that the tanneries were emitting too much effluent into the local river. The tanneries were ordered to install additional pretreatment equipment, and a new city waste treatment facility had to be built to bring discharge levels into compliance with the Clean Water Act. At this point UNITE entered into a formal alliance with the Tanners' Association and the local Chamber of Commerce to combat the new requirements, which they believed would threaten jobs. Towne acknowledges that at the time he was skeptical of the more extreme claims being made by employers, but that "my primary concern was to . . . save as many union members' jobs as possible" (personal communication, May 10, 1999). The Rainbow Alliance was the lead environmental advocate pressuring regulators to act, and again, it became the target of industry and union attack.

The main point of contention was the level of water quality that needed to be achieved in the Cayadutta Creek, a tributary of the Mohawk River, into which waste from the tanneries was dumped. The industry and the union took the position that, with the aid of some federal waste treatment grants, the applications for which were already in progress, they would be able to achieve what was designated by the DEC as a "Class D" level of water quality in the creek. They claimed that reaching the higher "Class C" level, which was being advocated by the environmentalists, would require excessive expenditures and result in financial ruin for the local tanning industry. Internally union and business leaders agreed that they could achieve the Class C requirements at a later point in time, but that plans to meet the weaker standard were already in motion, thus they should resist the reclassification of the creek for the time being. The Rainbow Alliance insisted that the creek might as well be reclassified at the higher standard, since the tanneries were already having to upgrade waste treatment equipment. This again led to heated conflict between the Rainbow Alliance and the union, including nearly violent confrontations between the chief adversaries. According to Sandy Fonda,

The Chamber of Commerce and the local union and some business folks contributed money to do a study proving that it would be financially disastrous for us to go ahead with this proposed reclassifying of the stream from a Class D to a Class C. And so we went to the press conference where they were releasing this. We got there and we were told that it was a private press conference. We were kicked out and Bill Towne, the union leader, slammed the door in my face. I was livid. (personal communication, May 10, 1999)

This classic "jobs versus the environment" dispute continued through the 1980s, polarizing the progressive community in Fulton County. However, several events took place that would soon alter the intermovement dynamics.

At the time of the Fulton County conflicts, unionists and environmentalists elsewhere in New York had established some cooperative ties. These relations grew out of the Love Canal disaster in Niagra Falls, New York, in 1978, during which over 200 families were forced to evacuate their homes because of toxic contamination of the land on which the homes were built. Local unions had lent support to the homeowners, among whom were several union members, thus establishing labor-environmental ties in the area. A major workplace environmental health incident in central New York brought unions and environmentalists in contact in that region, and together the ties in these two areas gave rise to the New York State Labor and Environment Network, a statewide network of unions and environmental organizations. Both Fonda and Towne had ties to people affiliated with the Network. Activists from this coalition began to encourage these two progressive leaders to try to resolve their differences. According to Fonda,

We were involved with Citizens' Environmental Coalition and the Labor and Environment Network, and they kept saying "Bill Towne comes to Albany to meetings on social justice issues, and he's really good and you should try to mend your fences." And they were saying to him "Sandy is really progressive, and if the two heads of these organizations can get together, maybe the two organizations can start working together." And we were saying "No Way! No Way!" (personal communication, May 10, 1999)

Under pressure from their colleagues, in 1989 both Fonda and Towne agreed to attend a workshop at a conference sponsored by the Labor and Environment Network in which they were asked simply to talk about the issues that they confronted in Gloversville. Fonda reported that "we fielded some pretty tough questions, me from the labor unions and Bill from the environmentalists." This interaction caused both of them to begin to rethink their relationship. The conference workshop was the first civil interaction between these two movement leaders, and the simple sharing of concerns in this forum allowed for greater understanding of one another's position in the dispute.

This meeting was followed by a chance encounter that would prove decisive in further altering the views of both actors. A year after the Network conference, Fonda and Towne were each independently appointed to the Board of Trustees of the Fulton-Montgomery Community College. According to Fonda, their duties as Trustees "forced us into a room together to work on something. We got talking a little bit, and we were both seeing the political sides of the college. Then we talked a little bit about political sides of the union-environmental thing. . . . We had some communication going" (personal communication, May 10, 1999). Issues at the college united Towne and Fonda against other board members and against the county government, which they charged was not adequately funding the school. In Towne's assessment, "It became really obvious to both of us that we could work together" (personal communication, May 10, 1999). These interactions allowed for crucial trust building between the two activists as well as a greater appreciation for one another's views on the issues that divided them. The evolution of Towne's thinking is evidenced by quotes in two news reports. In 1990 he was quoted in the local newspaper, the Schenectady *Daily Gazette*, regarding the conflict with environmentalists: "It's a matter of priorities. . . . The Rainbow Alliance is environment first, jobs second, and the union is jobs first and environment second" (Hammond 1990). Slightly over a year later he was quoted in another source regarding his relationship with Fonda: "It's always been, 'Is it jobs or the environment first?' . . . we've matured enough to where we recognize that it's both" (Goudreau 1991).

This transformation was manifest in increasingly cooperative relations between the union and the environmental community. In 1991 during some disputes between the Rainbow Alliance and one of the local tanneries, Colonial Tanning, regarding poor air quality, Towne arranged a meeting with Fonda and the tannery owner during which a compromise was reached. But over time Towne abandoned his role as mediator and was pushed further into an alliance with the environmentalists. The Rainbow Alliance and UNITE worked closely together against the North American Free Trade Agreement, successfully pressuring local lawmakers to oppose the measure in Congress. This experience strengthened their bond, and by this point, they considered themselves to be friends and allies.

A couple of years later the tannery effluent issue resurfaced. The industry had essentially won the first round of this dispute in the 1980s with the aid of organized labor. At that time the Cayadutta Creek retained its level D classification, as the industry and union had sought, and the new equipment waste treatment successfully brought the tanneries into environmental compliance with the required level of water purity. But by the mid-1990s, the DEC proposed raising the creeks' classification to the C level, the goal long sought by the environmental community. As might be expected, the tanneries objected and again cited the job loss threat to win union support for their position. This time, however, the union sided with the Rainbow Alliance. Towne recalled the agreement the union and the tanneries had made ten years earlier to accept the C classification at some future point:

All these guys, the Chamber, the tanners, are saying "we can't do it!" They were making the same argument. But at a public hearing in Johnstown . . . I read out loud what was said at the Chamber breakfasts ten years ago. I told them "This is what we did. Don't make this argument now. We said we'd do it and there's no reason not to." These cats never change their story. The industry, to a company, were all there claiming that they couldn't afford to do it. . . . For over fifteen years, they made those arguments about the Rainbow [being responsible for job loss]. Nobody listens to them anymore. The tanners will mouth them. There was a big brouhaha at the last hearing, but . . . nobody was listening to them anymore, certainly not our members, nor were the regulators. They, the industry, really went out of that last hearing humbled. (personal communication, May 10, 1999)

Fonda and Towne now work closely together. Workers in the plants report environmental violations to the union, and Towne passes the information on to the Rainbow Alliance. The Rainbow Alliance, in turn, uses its environmental expertise to help the union address problems at the plants. Knowing that UNITE and the Rainbow Alliance are working together, employers have asked Towne to use his influence to convince the environmentalists to drop lawsuits filed against them for environmental violations. Towne refuses these requests and instead directs them to clean up their practices. Regarding the new political alliance structure, Towne says, "We're seeing all this stuff come down. That information [regarding violations] comes in, it goes up to her [Fonda], she brings stuff in to us. We've got [the tannery owners] wired. They really can't move now, and they know it, because their worst fears have come true" (personal communication, May 10, 1999).

Except for the tannery owners, whose "worst fears have come true," the Fulton County story has a happy ending. Although jobs in the industry are still threatened by low-wage competition and labor-saving technology, employment is more secure for those currently employed because of the investment that has been made in the tanneries as a result of environmental requirements. The natural environment is now better protected, and the entire community, including the workers and their families, is enjoying those benefits. It is not difficult to imagine a far less desirable outcome, one in which labor-environmental conflict resulted in further environmental degradation, a compliant workforce fearful of job loss, and a fractured progressive community incapable of advancing any positive change. This has been the outcome in countless other communities around the nation in which the threat of job loss has led unions to side with employers against environmental measures. This conflict also plays out at the state and national levels, where measures supported by the environmental community are in many cases opposed by union leaders. These so-called jobs-versus-the-environment disputes have fractured progressive forces, preventing the implementation of government policies that are sensitive to both the environment and the needs of workers. A just and sustainable economy is the goal of both unions and environmentalists, yet still divisions between these two groups are common. The question that presents itself is why this is the case. Why did Bill Towne and Sandy Fonda consider each other to be enemies? And more importantly, how is this kind of division overcome?

This book is an attempt to answer those questions. Labor unions and environmental movement organizations are among the most powerful social movement sectors in the United States. When they act together, they can advance policies that protect both working people and the natural environment. Yet divisions between these two actors can yield environmental devastation and attacks on the interests of workers and their unions. The creation of a just and sustainable economy depends on the ability of these two social movement sectors to work together to advance this common goal.

Although the focus in this book is on alliances between labor unions and environmental organizations, an examination of these alliances can inform us about cooperation between social movement organizations (SMOs) in general. A deeper understanding of this type of cooperation is important, because a single organization is rarely capable of advancing any significant political goal by itself. Even when movement groups with similar interests work together, their effectiveness is generally limited. In the contemporary American context, the ability to enlist a range of allies in political struggles is often a crucial determinant of success; thus coalition formation has become an essential strategy for social movement organizations (Hula 1999; Steedly and Foley 1990; Thomas and Hrebenar 1994; Yandle 1983).

It is possible to identify some natural allies in political struggles, groups whose interests commonly converge. Most intergroup cooperation can be found within movement sectors. Environmental organizations, commonly work with other environmental organizations, and labor unions routinely act in concert with other unions. Similar organizations do experience some limited competition for resources and members, and on occasion major fissures erupt over these or other issues. But when it comes to advancing common political goals, cooperation is widespread within movement sectors.

Finding this type of convergence becomes less likely when we examine relations between different kinds of organizations. Some conflicts between organizations can easily be anticipated, as with, for example, interactions between a gay-rights organization and a conservative Christian group. Some alliances may also appear likely, as when an antinuclear organization joins with a peace group. But with other types of organizations, in which there are some overlapping and some conflicting interests, predicting intermovement ties becomes more difficult. In this book, the goal is to identify when cooperation is possible among these types of organizations.

Labor unions and environmental organizations represent an excellent example of this type of complicated relationship between organizations. As the case in Fulton County demonstrates, on occasion unions and environmentalists have utilized coalition strategies to further common aims. But this case also indicates that unions and environmentalists can find themselves on opposite sides of major struggles over economic and environmental policy. Conflict between labor unions and environmentalists has received a great deal of attention from academics and the media. Instances of cooperation between these movements, however, and when and how this cooperation arises have generated less interest.

An examination of recent history reveals mixed relations between environmentalists and labor unions. In some instances we see very close cooperation between unions and environmentalists. The so-called alliance of "Teamsters and Turtles" whose protests disrupted the meetings of the World Trade Organization (WTO) in Seattle in 1999 is the most well-known instance of labor-environmental cooperation in recent years (Greenhouse 1999; Moberg 2000). This coming together of a wide variety of unions and environmental advocates against the expansion of unrestricted trade was viewed by many as a surprising and unprecedented example of intermovement cooperation among workers and environmentalists. But there are many lesser-known examples of concerted efforts on the part of these two social movement sectors dating back several decades. For example, unions demonstrated early support for clean-air and clean-water legislation (Dewey 1998). In the 1980s national labor and environmental leaders facilitated the creation of state-based labor-environmental networks designed to challenge the antiregu-latory policies of the Reagan administration (Obach 1999). The WTO protests were actually preceded by other cooperative efforts to oppose the North American Free Trade Agreement and "fast-track" trade authority for President Bill Clinton. In other cases, joint efforts have been carried out to strengthen health and safety protections in the workplace and in the community, to promote recycling, and to mitigate the environmental damage caused by development (Gordon 1998; Gottlieb 1993; Obach 1999). Although the WTO protests were the first to capture national media attention, labor-environmental cooperation is in no way a new phenomenon.

Yet there are reasons why political analysts and the mainstream media expressed so much surprise over the labor-environmental cooperation in Seattle. In many cases the ties between these two movements are tenuous. In 2002, just a few years after Seattle, sharp divisions emerged between environmentalists and the Teamsters over a Bush administration proposal to drill for oil in the Alaskan National Wildlife Refuge (ANWR). That same year the United Auto Workers (UAW) also had a falling out with environmentalists over efforts to raise the federal Corporate Average Fuel Economy (CAFE) standards, part of the move to reduce greenhouse gas emissions. This followed previous disputes over the Kyoto Protocol, an international treaty designed to combat global warming. Despite examples of intermittent collaboration, there has also been a great deal of conflict between environmental and union interests over the decades (Audley 1995; Buchanan and Scoppettuolo 1997; Dreiling 1997; Dunk 1994; Foster 1993; Judge 1995; Larsen 1995; Logan and Nelkin 1980; Ramos 1995).

But workers are not typically the lead opponents of environmental measures. Although UNITE played a major role in the conflict in Fulton County, it was the tanning industry that led the campaign against the environmental proposals. Environmental movement organizations are most commonly pitted against private-industry executives who wish to avoid the costs and constraints of environmental regulation (Arrandale 1994; Ringquist 1995; Ruben 1992; Sanchez 1996). It is when industry seeks allies in opposition to environmental measures that workers are drawn into the fray. Employers seek to enlist workers to rally against environmental measures by using the threat that losses or decreases in profit suffered as a result of implementation of such measures may result in layoffs or a complete shutdown. Knowing that a threat to corporate profits will not move the public, a more sympathetic victim is necessary to win public support, and workers are the obvious group to serve this purpose. Industry opponents of environmental measures will typically fade into the background and carry on low-profile lobbying while workers are presented as the public face of environmental opposition. Often cast as issues of jobs versus the environment, these conflicts have captured the most attention and have helped to shape the perception that environmental protection is antithetical to economic expansion, job preservation, and the interests of workers generally (Cooper 1992; Goodstein 1999; Gray 1995; Kazis and Grossman 1991).

Virtually every instance of labor-environmental conflict involves either a threat to existing jobs or the loss of potential jobs. The UAW feared that unionized American automakers would lose market share and eliminate jobs if higher fuel economy standards were imposed on the sport utility vehicles they manufactured. In the case of ANWR, the Teamsters saw the prospect of job creation with the introduction of oil drilling in the Alaskan Refuge. In a well-known case of labor-environmental conflict from the 1980s and 1990s, timber industry workers in the Pacific Northwest feared job loss if measures were imposed to save a threatened species of owl. Other labor-environmental conflicts have erupted over issues such as nuclear energy, bottle bills, development restrictions, and toxic use reduction (Buchanan and Scoppettuolo 1997; Judge 1995; Larsen 1995; Logan and Nelkin 1980; Smith 1980). In all of these cases, workers and their union representatives perceived environmental measures as posing a threat to jobs.

Although each issue has unique elements that may unite or divide segments of the labor and environmental communities, to better understand patterns in these relationships, it is first useful to examine how these two movements are situated in the larger political economy of the United States. The structure of the U.S. capitalist system generates contradictory drives toward private material gain and the protection of the public from the negative externalities that such economic activity often entails. This economic system not only facilitates the pursuit of profit by private firms but compels most individual workers to seek personal security through private employment. In this way the interests of workers become tied to those of their employers, at least in terms of preserving the enterprise that provides profit for one and wages for the other. The tannery owners in Fulton County and the workers at those plants both have an interest in maintaining the local tanning industry. This is true despite the adversarial relationship that may exist regarding the distribution of resources within the firm or the control of the labor process.

In the pursuit of private gain and security through the operation of an enterprise, certain effects may "spill over" onto the public. As firms seek to increase profits by externalizing costs, the public is made to bear the brunt of their actions. Pollution serves as the classic example of such "negative externalities." The tannery owner may save money and increase profits by dumping toxic waste into the river, but those living downstream and the public in general pay the price. Groups organized around protecting the public interest, such as environmental advocates, may then find themselves in conflict with the private interests that are threatened by measures that will interfere with their activities. Herein lies the root of labor-environmental conflict: an economic system in which private control over material resources and the pursuit of individual gain generate costs for outside parties and the public at large.

It is the structure of the capitalist economy that creates the conditions for conflict between private economic interests and the protection of public goods such as the environment, but it is the structure of American democracy that determines how such tensions are translated into political disputes. Following independence, the nation's founders were wary of centralized power and created a system they thought would prevent factions from gaining undue control. The mechanisms they put into place, designed to prevent the centralization of power, have the effect of generating multiple competing interest organizations (Wilson 1993). Yet with some notable exceptions, certain elite factions managed to maintain relatively tight control over policy and lawmaking for much of the nation's history despite these mechanisms. Unions and some other organized constituencies achieved some power in the first half of the twentieth century, but it was not until the 1960s that other popular sectors began to exercise greater influence. Political and economic conditions during this period spurred broad popular mobilization that gave rise to a proliferation of movement organizations representing a wide cross section of the population (Berry 1984; Salisbury 1990; Schlozman and Tierney 1986; Walker 1983).

Government policies and other institutional mechanisms shaped the types of organizations that emerged out of this mobilization. The creation of policy and the associated bureaucratic structures influenced how interest organizations operated and even shaped the way in which interests were defined. The creation of public agencies designed specifically to address environmental issues fostered the formation of movement organizations that took environmental matters as their central concern. Previously, worker mobilization had given rise to government policies and bureaucratic structures that channeled union interests toward specific workplace issues. The fact that Bill Towne and UNITE prioritized jobs over the environment is in part a reflection of the workers' preference ordering, but it is also a function of the legal and organizational structure of the labor movement in the United States, which facilitates a focus on wages, hours, and working conditions in preference to more general concerns such as environmental protection.

The organizational form of the emergent movements and the development of government mechanisms designed to address specific issues, in turn, have had important implications for how these organizations relate to one another. And the political space, now crowded with a multitude of movement organizations, *necessitated* these interorganizational relationships. This is especially true given that political parties have largely abandoned their role of mobilizing different constituencies and synthesizing interests into broadly acceptable policies. These intermovement ties have proven to be crucial, because in light of the more fragmented political landscape, caused by narrowly focused issue organizations combined with the declining significance of political parties, these relatively narrow interests must work together in order to achieve their goals. It is these developments that have caused social movement coalition formation to become a mainstay of politics in the United States (Hula 1999; Jenkins-Smith and Sabatier 1994; Keller 1982; Loomis 1986; Moe 1980; Sabatier 1988). Thus the market economy sets the stage for conflict between private interests, such as capitalists and those they employ, and public-interest organizations, such as environmental groups, and the legal and political structure of American democracy facilitates the form that organized advocates will take. Within this macro-political-economic context, we can begin to analyze labor-environmental relations and why cooperation between these two sectors is so important, yet often so elusive.

The need for coalition formation is greatest for groups that are disadvantaged in terms of resources and political influence. The pooling of resources and political support to advance one's goals more efficiently is the central advantage of coalition formation as a strategy (Chertkoff 1970; Gamson 1961; Kelley 1970; Kelley and Arrowood 1960; Leiserson 1970; Von Neumann and Morgenstern 1947). Organized labor and environmental organizations, although among the most powerful popularly based movements in the country, are, in terms of resources, at a disadvantage relative to corporate interests, which often act as their political adversaries.

But labor's allegiance within this triad is difficult to discern. Unions can be seen as situated between environmentalists and their employers when environmental issues arise. They share a common interest with employers in maintaining enterprises and the jobs that they provide to the unions' members; thus they are susceptible to any perceived threat to employment. But the employers for whom their members work are also the adversary they were created to combat. Bill Towne worked with the Fulton County tannery owners against the threatened environmental measures, but the union and the owners were bitter enemies on many other issues. Conflict around workplace issues has the potential to expand into other realms, pushing workers into alliance with anyone who shares the same adversary.

Beyond this traditional adverserial relationship between workers and owners, workplace safety issues and community environmental concerns make environmentalists workers' natural allies in the effort to restrain employer abuses. The hazardous substances that the Rainbow Alliance wanted to restrict are the very ones that the tannery workers were exposed to on a daily basis. The community that the alliance was seeking to protect is also the community in which workers and their families live. As demonstrated by the Fulton County case, unions have entered into environmental policy disputes on both sides. When such a dispute arises, they become the target of appeals from both employers and environmentalists, and given the importance of coalition support as a determinant of movement success, their leanings can determine the policy outcome (Steedly and Foley 1990). When unions do enter into such disputes on behalf of the environmental cause, they can often provide the crucial backing necessary for environmental policies to be passed over the objection of industry opponents. Although it is less common, on some occasions environmentalists also offer crucial support to workers in disputes with their employers (Gordon 1998; Rose 2000). Situations in which these groups do not align with one another, however, often yield policy outcomes that benefit industry at the expense of workers, the natural environment, or both.

Thus when and under what circumstances unions and environmental advocates are able to align is an important question for understanding the outcome of many policy disputes. Perceptions about the particular repercussions of a given measure can determine whether various unions will side with their employers or join in coalition with the environmental community. But beyond the question of specific short-term instances of interest alignment, a more important question is the overall quality of the relations that exist between these two movement sectors. The more general relationships between labor unions and environmental groups can have a broader effect on how these organizations will behave when specific issues arise. Cooperative ties, once established, can shape the way all future issues are addressed. Thus, beyond any specific issue, the general quality of the intermovement relations that are established can have implications for the overall direction of political and social change.

When unions and environmentalists have positive relations with one another and form an ongoing alliance, they present a formidable political force potentially capable of redirecting economic and environmental policy in fundamental ways. The significance of the ties that result from these relations can be seen in the political strategies used by political parties at the national level. Environmentalists and unions both represent important elements of the Democratic Party base, and in many cases Democratic officials struggle to avoid alienating one or the other when a policy has contradictory implications for workers and the environment. Republican strategists often take advantage of the potential divisions that arise in such instances and attempt to peel away Democratic support by fostering dissension between these crucial segments using particularly divisive issues (Kahn 2001). Thus the cohesiveness of these two movement sectors can be crucially important beyond any single issue. Alignments of their interests around particular issues and the quality of their ongoing relations are the central concerns of this study.

The macro political and economic structures create the context in which organizations, such as unions and environmental SMOs, form and seek to advance their objectives. We must also focus, however, on the middle level to see how organizations behave within that context to fully understand when intermovement alliances will emerge. When examining political-alliance formation, most scholars focus on the intersection or divergence of the interests of the actors involved (Gamson 1961; Riker 1962). In the conventional approach, organizations are viewed as rational systems oriented toward the achievement of specific goals, and they choose to ally themselves with other organizations when doing so will enable them to achieve their objectives (Scott 1992; Simon 1976). Based on this view, one might argue that unionists will oppose environmental policies when such measures present a threat to jobs or the general economic well-being of workers, and they will support environmental policies when they advance these core interests. Although such a hypothesis offers a good framework for beginning our analysis, it readily becomes clear that the question of when the interests of workers are threatened is a very complex one, and that different organizations are likely to have different interpretations regarding this question. Thus when considering the actual process of intermovement relationship development, we must more closely examine the question of interests in addition to an array of organizational issues that emerge.

As we do so, we must first recognize that the American labor movement has a very diverse and complex structure. In addition to parallel union and union federation organizational hierarchies, at times individual unions cover workers with very different jobs at different employment sites or even within the same site, depending on how the state governing body defines a particular bargaining unit. This is important because different employment sectors are affected in different ways by particular environmental policies. With regard to any given policy proposed, some workers may see their interests as being threatened, whereas others will identify a potential benefit, and still others may perceive themselves as not affected at all. Certain employment sectors are more obviously threatened by environmental measures than others. Workers in extractive industries, for example, may clash with environmentalists seeking wilderness protection. On the other hand, some employment sectors are clearly in a position to benefit from certain environmental policies. For example, public employees who work for a regulatory agency may see an employment benefit to greater environmental protection. Thus we must first dissect how distinct employment sectors are affected by different environmental policies before we can make predictions about likely alliances between labor and environmental groups.

But in addition to identifying the discrete interests of different sectors of the labor movement, we must also examine how those interests are specified by individual unions. Although legal and organizational structures foster common tendencies within similarly situated groups, there is still great diversity in terms of how unions behave politically. Some unions are inclined to oppose any perceived threat to the employment interests of their particular members, whereas others take a broader approach to interests, factoring the overall well-being of the working class into their strategic analysis. Some emphasize economic issues, whereas others give more weight to issues such as the health and safety of their members, their own concern with the protection of the natural environment, their common cause with environmentalists in fighting employers and regulating industry, and mutual support for certain political candidates.

Similar variation can be found among the organizations that make up the environmental movement, a movement that pursues diverse goals and is more decentralized and organizationally complex than the labor movement. Some environmental organizations seek to promote policies exclusively designed to protect the natural environment, yet others are primarily concerned about the threats to human health and safety posed by environmental degradation. Both types of groups would be considered "environmental" organizations, but the contrast in the goals they seek to advance will make some more likely than others to ally themselves with organized labor. Although it is still possible to begin with a rational-systems framework in which organizations methodically pursue specified goals, these varying orientations among groups within the environmental movement must be taken into consideration when one is conducting a conventional coalition analysis based on the interests of the actors involved.

Examining the core interests that underlie the relevant movements, however diverse and complex these interests are, is an important first step in understanding when labor-environmental alliances will form. Analysis of the way in which the policies advocated by one group affect the members of another is an important issue. Although much coalition behavior can be understood in these terms, such a one-dimensional interest approach fails to address the complexities involved in interest formation and political strategizing. As suggested above, unions represent worker interests in a broad array of areas, pushing them in several directions simultaneously. The interests upon which unions base their actions or develop their priorities are in no way given, allowing for a great deal of complexity even if immediate member interests were the sole determinant of union action. But an analysis of intermovement relations based upon a pure interest assessment is further complicated by the fact that the goals of environmental organizations are difficult to classify in terms of interests.

Some theorists argue that a fundamentally different logic underlies organizations from these distinct movements (Larana, Johnston, and Gusfield 1994; Melucci 1980; Offe 1987). These "New Social Movement" theorists argue that the movements that have developed since the 1960s, such as the environmental or antinuclear movements, are "values" centered as opposed to being based on the achievement of private interests. Labor unions, some argue, can be easily understood within an interest framework. They are designed to protect and advance the interests of their members, particularly as they relate to wages, hours, and working conditions. New Social Movement theorists argue that this type of interest-based organization was the dominant form prior to the 1960s. Since the 1960s, however, we have seen the rise of social-movement organizations that are not focused on the particular interests of their members, but rather are oriented toward achieving public goods or toward advancing values that have no immediate personal benefit to the membership. For example, whereas a negotiated wage increase is clearly in the interest of each individual member of the union that negotiates it, the preservation of an endangered species may have no direct impact on an individual member of an environmental group that supports such a cause. Since advancing personal interest is not the primary goal of these organizations, New Social Movement theorists focus on the values component of mobilization. Individuals join or contribute to such groups not because they hope to gain personally from the work of the organization, but as an expression of their values. In this context, the rational strategic pursuit of organizational goals is complicated by the expressive-value element of environmental activism. Understanding the alignment of interests among two types of organizations is difficult if one type is not based on the pursuit of interests at all.

Some have criticized New Social Movement theorists for drawing contrasts too sharply between the interest-based "old social movements" and the values orientation of the new (Larana, Johnston, and Gusfield 1994). Both types of movements can be seen as operating on the basis of some combination of values and interests. Furthermore, once the policy positions of labor and environmental organizations are identified (whether they are based on values or interests), it is still possible to analyze the extent to which the respective goals of the organizations align or diverge. But beyond questions regarding the foundation of their goals, some important differences do exist in terms of the structure and orientation of these two movement sectors. For example, resource mobilization theory suggests that obtaining funds is a crucial determinant of organizational survival and success (McCarthy and Zald 1977). In looking at the differences between new and old social movements, one must take into consideration how such voluntary, values-oriented organizations like environmental groups attract and maintain support and identify how this may place restrictions on the political strategies they adopt, including their capacity to forge intermovement alliances. Environmental leaders, like Sandy Fonda, must constantly consider how the position taken by the organization will be received by the members. Would cooperation with a union on a given issue be seen as a betrayal of the environmental cause? Would a job-saving compromise in a particular situation cause membership to drop off, thus threatening the very survival of the organization?

Labor unions, whose members are often automatically affiliated with the union through their workplace, do not face the same constraints in terms of mobilizing resources. Yet unions face different obligations in terms of achieving private material gains for their members. Bill Towne had to consider the interests of his members when devising a political strategy for tackling the environmental issues that had arisen regarding the Fulton County tanneries. But for him the immediate threat was not dissatisfied workers' quitting the union, but rather, the union's collapsing because of the loss of jobs. Again, these constraints on unions' actions will shape the political strategies available to union leaders and either induce or inhibit coalition formation with environmental SMOs. Thus, the political actions taken and strategies adopted by these new- and old-movement organizations are shaped and constrained in very different ways, further complicating the question of when these organizations will be brought into alliances with one another.

There is yet a third level of analysis that must be considered when we are attempting to specify coalition practices. In addition to the macro-political and economic conditions that provide a framework within which movement organizations operate and the meso-level organizational characteristics that shape their behavior, for coalitions to form, actual *individuals* have to interact in a concerted manner to establish relations with one another. Bill Towne and Sandy Fonda would probably not have formed their alliance had they not been brought together by their personal ties with people involved in the New York State Labor and Environment Network. Even their chance appointment as trustees of Fulton-Montgomery Community College was crucial in their case for facilitating the interpersonal bonds that are decisive for the maintenance of organizational ties. Thus a full understanding of intermovement relations must also consider this micro level of analysis.

Some theorists have noted that labor and environmental movement participants face particular difficulties in regard to the establishment of interpersonal ties because of the cultural differences that divide members of these two movements (Eder 1993; Rose 2000). Middle-class professionals dominate the environmental movement, and they are embedded in a different cultural milieu than blue-collar workers. The views of members of these two groups on political action, the nature of work, organizational functioning, the basis of knowledge, and the role of nature differ in some important ways, complicating efforts by individuals from these movement sectors to achieve the common understandings necessary for them to work together. The racial and ethnic differences between a predominantly white environmental movement and an increasingly minority-dominated labor movement add another level of complexity.

Many factors shape the extent to which labor unions and environmental organizations are likely to engage in cooperation or conflict. Broad political and economic structures create the macro context in which movement organizations form and the goals toward which they are oriented. Meso-organizational level variables also come into play when we examine movement groups that are organized on the basis of very different principles with distinct intramovement ties and organizational maintenance needs. Lastly we must not ignore microlevel considerations, such as the interpersonal ties between particular individuals within these movements. Each of these factors is examined in detail in the chapters that follow. Although barriers to labor-environmental cooperation exist at every level, in many instances unions and environmentalists transcend those barriers and find that they are able to work together to create a more just and sustainable society. The goal in this book is to identify exactly how such cooperation is developed and maintained.

To uncover the conditions that facilitate or inhibit intermovement cooperation between unions and environmentalists and the process by which such cooperation is created, I examine several cases of labor-environmental interaction. Interviews with dozens of labor and environmental leaders and activists have been used to piece together how these intermovement relationships develop. Cases drawn from states around the United States provide examples of successful labor-environmental cooperation and of heated conflict. The chapters are organized not around single instances of intermovement ties, but around the common issues and themes that have emerged across cases. What is of importance is not the detail associated with any particular instance of labor-environmental conflict or cooperation, but rather the common challenges that emerge in all such cases. The themes explored address each of those introduced above, from the macrostructural conditions that give rise to social movement organizations oriented around a limited set of political objectives down to the microlevel interactions of individuals rooted in distinct movement sectors. Below I provide a chapter-by-chapter preview of the issues analyzed in the book.

Most coalition theory begins with assumptions regarding the way in which organizations act on the interests they are designed to advance. Thus it is important to first consider how different segments of the population are affected by environmental policies. Who benefits and who, if anyone, loses as a result of environmental protection? Is the jobs-versus-the-environment trade-off a myth or a frightening reality for workers? If there is an economic downside to environmental policy, who pays the price? Can this question be answered in class terms, with the largely middle-class environmental movement well insulated from the economic repercussions of environmental policy, forcing economic costs on a vulnerable working class? Or is this issue more complicated, such that only certain employment sectors ever experience the economic downside of environmental policy? What about interests that go beyond economics, like health or political power? Are these win-win issues for workers and environmentalists? In chapter 2 all of these questions are explored. Using the results of economic analyses, the interest configuration of environmentalists, workers, and their employers is examined. This provides a rough outline of which groups are likely to come into conflict and which stand to benefit from common policies.

Although an examination of interests provides a natural starting place for an analysis of interorganizational alignment, what is more important is how those interests are actually acted upon in the political sphere. In chapter 3 I provide a historical overview of labor-environmental relations. Relations between labor and environmental organizations began on a positive note at the start of the contemporary environmental movement in the early 1970s, but soon economic concerns generated skepticism about environmental measures within a labor movement facing recession, job loss, and rapidly declining membership. Over the next three decades we can see the emergence of jobs-versus-the-environment struggles punctuated by concerted efforts to build labor-environmental unity. The 1970s and 1980s included their share of conflict over such issues as nuclear power and forest preservation. But there were also major success stories in these decades, such as the cooperative effort to gain access to information regarding the use of toxic substances in the name of worker and community health and safety. During the 1990s free trade emerged as a unifying issue among unions and some segments of the environmental movement. Yet despite substantial cooperation and the high-profile unity seen in the streets of Seattle, issues related to global warming as well as other issues continue to drive a wedge between environmentalists and important segments of the labor movement. The waves of cooperation and conflict and the variable relationships that have developed between different unions and environmental organizations demonstrate the complexity of interactions between the two. Economic and political interests certainly play a role in generating conflict or cooperation, depending on the strategic advantages to be had under different societal conditions. But as with the interest analysis offered in chapter 2, it is clear that the quality of labor-environmental ties cannot be reduced to a simple interest configuration. Because an interest assessment alone cannot reveal the type of labor-environmental relations that will emerge, more in depth analysis of the *process* of intermovement relationship development is needed. The subsequent chapters are all dedicated to examining the important elements of that process.

The case studies used for the research presented in this book are drawn from five states: Maine, New Jersey, New York, Washington, and Wisconsin. States were used as the initial focus of the analysis because they provide a sound basis for comparison. Each state has its own set of environmental organizations and a centrally coordinated state labor federation. State environmental and economic policies then serve as the basis for labor-environmental cooperation or conflict. Chapter 4 provides information on the relevant economic and political conditions in each of these five states. In addition I identify the main labor and environmental actors in the five states and provide an overview of the general quality of labor-environmental relations.

In chapter 5 I examine the organizational constraints that limit the ability of unions and environmental groups to engage in coalition activity. I first ground the analysis in the context of American politics to demonstrate the way in which the U.S. political structure tends to give rise to numerous, narrowly focused movement organizations. I then argue that this crowded political field and the limited set of issues addressed by each organization creates an organizational dilemma when coalition formation as a strategy is considered. Organizational maintenance needs require that an SMO remain able to distinguish itself from others. Yet such distinctiveness reduces issue overlap with other organizations and inhibits the potential for coalition work, a dilemma that I refer to as the "coalition contradiction." This condition afflicts SMOs in different ways depending on their "organizational range" (the number of issues that a group seeks to address) and the type of goals they strive to achieve. The need of unions to advance the material interests of their members creates some barriers to intermovement cooperation; voluntary new social movement organizations, however, like environmental groups face a more acute strategic dilemma. In chapter 5 I analyze the relationship between the structural conditions that give rise to different types of movement organizations and the way in which distinct organizational characteristics limit coalition participation.

The structural conditions described in chapter 5 shape the propensity for movement organizations to engage in intermovement coalitions, but those conditions are not a determining force for coalition involvement. Organizations are capable of self-transformation. Both external conditions and internal processes can alter organizations in ways that change their coalition propensities. In chapter 6 I analyze these changes in terms of "organizational learning." Changes that occurred within some of the labor and environmental organizations examined here reveal two types of organizational learning: experiential learning and learning through interaction with others. Experiential learning occurs when organizations face crises or fail to achieve organizational goals and, through a process of trial and error, develop new strategies and, on occasion, incorporate new goals as a result of their experience. This expansion of organizational range then widens the field of prospective coalition partners. The second type of learning, learning through interaction with others, typically results from ongoing contact with other movement organizations in a context of mutual effort. Such interaction builds trust and facilitates the homogenization of goals among organizational leaders, thus allowing for still more cooperation in a broader range of issue areas. Organizational learning and range expansion can allow for more coalition participation, but the cases indicate that organizational learning is durable only if certain conditions are met.

In chapter 7 the process of cross-movement interaction is explored in greater depth. In particular, attention is focused on the role of "coalition brokers." Brokers are the actors who bring movement organizations together in an effort to foster cooperation. They typically occupy a position that bridges the divide between the distinct groups allowing them to communicate with both sides and to frame issues in ways that resonate with both constituencies. In addition to the role played by these individuals, certain organizations are also commonly involved in the broker role, including citizens' groups and occupational health advocates. In this chapter I document the way in which coalition brokers utilize bridge issues to bring unions and environmentalists together.

Chapter 8 addresses the question of the cultural divide between the typically middle-class participants in the environmental movement and the working-class members of the labor movement. Although the constituents of the two types of organizations are less distinct in terms of class than they once were, some theorists argue that differences in class culture inhibit cooperation between these two movement sectors. In chapter 8 the extent of those cultural differences is gauged. I argue that although some cultural differences do exist between some unionists and environmental advocates, cultural distinctions pose only a minor barrier to intermovement cooperation. I also argue that the cultural gap identified by others is better understood as a manifestation of organizational differences rooted in legal and structural pressures as opposed to class culture.

Taken together, each of the book's chapters represents one piece in a larger puzzle. The central issue is to understand when and under what circumstances we can expect cooperation between organized labor and the environmental movement. But the conclusions reached also have more general application. Macropolitical and macroeconomic structures can be seen as the foundation of the social-movement politics that characterize American democracy. A core-interest analysis allows us to weigh the effect of those factors taken by many to be the beginning and the end of political-coalition study. But the examination in this book goes beyond that to consider the impact of organizational and microlevel variables as well. Consideration of environmental SMOs and labor unions on the organizational level enables us to better understand the strategic constraints placed upon voluntary purposive organizations relative to those primarily organized around private material interest. In addition to the structural forces that shape political outcomes at the macro- and the meso-organizational levels, the microlevel interactions between individuals must also be factored into the analysis. This is where we consider the basic mechanics of organizational contact and the influence it has on key actors, allowing individual bonds to grow into movement alliances. In the concluding chapter I assemble the pieces to suggest an overall framework for coalition study. Although not a completed picture, it provides a basic outline for understanding not only labor-environmental relations, but also intermovement alliances in general.

A Note on Terminology

This book examines the relationships between labor unions and environmental organizations. The full range of relationships will be considered from intense conflict to close cooperation. When labor-environmental relations are at their best, they may involve a "coalition" among various unions and environmental groups. Scholars have used the term "coalition" in many different ways. The most basic definition is perhaps that offered by William Gamson: "A *coalition* is the joint use of resources by two or more social units" (1961, 374). Although the term "coalition" always implies the coalescing effort of more than one actor toward a particular goal, the level of coordination can vary dramatically from coalition to coalition. Political scientists often refer to "electoral coalitions" when analyzing the broad array of interests and organizations that support a given candidate or party. However, this does not necessarily imply any coordinated action or formal recognition of one another on the part of the actors involved (Axelrod 1986). Indeed, there may, in fact, be a large degree of disagreement and conflict among the actors involved in such a "coalition." For example, a Democratic presidential coalition may include environmentalists, labor unions, and high-tech corporations, despite their divergent views on a wide range of issues and a total lack of strategic coordination among them. However, "coalition" may also be used to refer to coordinated action among organizations. This can take the form of one-time, short-term, issue-specific coalitions that may involve nothing more than adding an organization's name to a list of supporters. Or it can involve highly coordinated efforts among groups that form stable, long-standing alliances that work together on a number of issues. It can even include the creation of a separate office and staff and the dedication of other resources to overseeing coalition activity. For my purposes I use "coalition" to refer to actual coordination among labor and environmental organizations, not the electoral type of coalition that implies only common support for a candidate or issue. The full range of activity among coalitions will be considered, from those that involve limited effort on a single issue to ongoing partnerships. I use the term "coalition" frequently, but given the range of relationships that it can imply, a more thorough description will be provided as needed for any coalitions referred to in this book.

2

Interests and Alliances: Economic and Political Determinants of Labor-Environmental Relations

An appropriate starting point for an investigation into labor-environmental cooperation or conflict is with the assumptions central to virtually all political and coalition theory. That is, that organizations created to advance certain interests will cooperate with one another when their interests align and mutual gain from the cooperation is anticipated, and that they will come into conflict when the goals of one threaten the interests of the other (Bacharach and Lawler 1980; Chertkoff 1970, 1976; Gamson 1961; Hinckley 1981; Hojnacki 1997; Hula 1999; Lawler and Young 1975; Leiserson 1970; Loomis 1986; Polsby 1978; Riker 1962; Wright and Goldberg 1985). With respect specifically to labor-environmental relations, most attention has been focused on the threats posed to the economic interests of workers by environmental protection (Burton 1986; Buttel 1986; Dewey 1998; Dreiling 1998; Gladwin 1980; Gordon 1998; Gottlieb 1993; Jones and Dunlap 1992; Kazis and Grossman 1991; Morrison 1986; Potier 1986; Ringquist 1995; Siegmann 1986; Watts 1986). Although the popular media often present this as a simple issue of trade-offs—jobs versus the environment (Cooper 1992; Goodstein 1999; Gray 1995)—more sophisticated treatments of this matter have found the question to be far more complex.

Under certain circumstances some employment sectors may be harmed by environmental regulations, whereas in other cases such regulation generates economic benefits. The appropriate question, then, is, who benefits and who is likely to suffer economically as a result of environmental regulation? Scholarly research on this dimension of the topic can be broken down into two general categories, broad, class-based theoretical analyses and more empirical economic analyses of the effects of environmental policy on the economy. Another relevant area of concern is that of the political interests of the organizations within the labor and environmental movements. Although the issues that separate or divide unions and environmentalists often involve questions regarding the economic effect of environmental policy, both movement sectors have a range of political goals that do not bear directly on one another. Both therefore have political and electoral interests that may converge for reasons unrelated to any particular policy. For this reason political and electoral interests must be examined independently, apart from specific policy issues.

In this chapter the research on economic interests as they relate to environmental policy will be assessed to provide a general sense of the distributive impacts of environmental regulation. Union and environmental political and electoral goals will also be examined to determine whether the electoral realm creates tendencies toward labor-environmental cooperation or conflict. This general assessment of interests and the way in which they overlap or diverge will provide a foundation from which we can then explore the actual process of intermovement relationship development. As will be seen in later chapters, interests alone do not determine the quality of intermovement relations. Nonetheless, they serve as a necessary starting point for understanding how groups within the labor and environmental sectors may be inclined toward cooperation or conflict.

Theoretical Considerations of Class and Environment

Marx (1967) predicted that the ongoing development of a capitalist economy would result in a growing schism between continually concentrated capital and an increasingly immiserated working class. Although there was ample evidence for such a trend, especially in the first decades of the industrial revolution, later developments contradicted some of his predictions. In particular, he did not foresee the growth of the professional middle class that occurred in many advanced capitalist nations. How to understand the interests of this class in relation to the working class has presented a puzzle for class theorists (Eder 1993; Gouldner 1979; Wright 1976). Achieving such an understanding is especially difficult when attempting to assess the configuration of these interests in regard to specific issues, such as environmental policy. A number of theorists, however, have taken up the question of how the costs and benefits of environmental protection are distributed among the classes, thus indicating the relationships that should be expected among them.

It has been established that lower-income groups suffer disproportionately from the effects of environmental degradation in terms of its negative health consequences and other quality-of-life issues (Baumol and Oates 1979; Berry 1977; Bryant 1995; Bullard 1993; Lester, Allen, and Hill 2001; United Church of Christ Commission for Racial Justice 1987). Research has demonstrated that, sometimes as a matter of policy, hazardous, environmentally undesirable facilities are sited in or near low-income communities (Bullard 1993; Field 1998). The health implications for communities surrounding such facilities are well established (Head 1995).

Yet some have argued that in addition to the disproportionate share of the harmful effects of environmental degradation that the working class endures, policies designed to *protect* the natural environment also tend to impose a greater economic burden on the working class (Buttel 1975; Buttel, Geisler, and Wiswall 1984; Buttel and Larson 1980; Morrison, Hornback, and Warner 1972; Zimmerman 1986). Whereas white-collar middle-class workers are in many ways shielded from the direct economic impact of environmental policy, blue-collar workers, particularly in the industrial and extractive industry sectors, face job loss and economic uncertainty if environmental policies threaten the profitability of their employers. Opponents of environmental measures often highlight class differences when seeking to mobilize blue-collar workers. Middle-class professionals, who are disproportionately represented within the environmental movement, are depicted in these disputes as insensitive to the plight of blue-collar working people (Brown 1995; Foster 1993). The jobs-versus-the-environment conflict is presented as one in which blue-collar workers will be made to suffer for the benefit of the environment and its primarily middle-class advocates (Cooper 1992; Gray 1995).

This scenario can explain the "growth coalition" of labor and capital, which in some cases has opposed environmental measures, and their middle-class sponsors, that they believe threaten the economy. According to Allen Schnaiberg (1980), labor and capital both stand to gain economically from continually expanded production. In Schnaiberg's assessment competition within market economies fuels a "treadmill of production" in which social surplus must be channeled into greater productivity to increase profits and maintain a given level of socioeconomic welfare. At the same time this treadmill yields increased material withdrawals in the form of resource extraction and additions in the form of pollution, which together represent the destruction of the natural environment.

Based on Schnaiberg's analysis, workers and their union representatives adopt a strategy of "economism" in which they support expanded production and focus their energies on attaining a greater share of the surplus generated, while neglecting environmental concerns and opposing those who promote them. Although white-collar middle-class workers can be seen as ultimately dependent on economic expansion as well, their occupations are generally farther removed from production practices that directly harm the natural environment than those of blue-collar workers. According to some theoretical analyses, this three-class configuration is likely to yield a labor-capital alliance against a more environmentally inclined middle class. In the words of Claus Offe (1987), labor and capital are inclined to unite in opposition to environmentalists and other middle-class movement actors who represent "the needs and values of those who neither contribute to industrial production nor conform to its values and standards of rationality" (99).

Many of those examining the environmental issue have suggested that one form of class coalition or another will inevitably emerge from this array of interests (Gladwin 1980; Morrison 1986), and although much of this analysis suggests a labor-capital alliance against middle-class environmentalists, consideration has also been given to the possibility of a broad labor-environmental coalition (Buttel and Larson 1980; Kazis and Grossman 1991; McClure 1992; Miller 1980; Offe 1987). There are several grounds for suggesting that this is the more likely alliance to emerge. As stated previously, the working class bears a disproportionate share of the harm due to environmental destruction, giving working-class people a clear interest in environmental protection. In addition, survey research has demonstrated that environmental concerns are widespread among the working class (Jones and Dunlap 1992; Van and Dunlap 1980) and that lower-income people are actually more willing than others to sacrifice economic expansion in favor of environmental protection (*Gallup Poll Monthly* 1995). The natural antagonism between workers and employers could also motivate worker organizations to ally themselves with others who challenge corporate interests. Evidence of this type of alliance can be seen in the tendency for European social democratic and labor parties to take up environmental causes or to ally themselves with Green parties. The tendency for both unions and environmentalists to support the Democratic Party in the United States offers further evidence of this type of alignment.

Such broad, class-based theoretical assessments may, however, oversimplify the political coalitions that actually emerge around environmental issues. Some have argued that given the array of interests within social-class categories, no broad class alliances can be anticipated in regard to the environment, and that only ad hoc coalitions around particular issues will emerge (Watts 1986). Frederick Buttel and Oscar Larson (1980) have argued that no class as a whole has a distinct interest in environmental protection:

since the benefits secured by environmentalists are distributed so broadly, environmentalism has no natural constituency among the enduring social categories. Instead, the rooting of environmental concerns in key production and consumption institutions, especially in terms of the ways that environmental controls would affect various groups of producing organizations and consumers, dictates that various groups will adopt an array of environmentally related interests and ideologies. (326)

As will be demonstrated this array of environmentally related interests and ideologies has given rise to labor-environmental conflict and cooperation in different instances. To better assess the complex relations that result, some have argued in favor of a new class scheme in which class definitions based on the relations of production are supplemented by "environmental classes." Using this approach, fissures along class lines can be understood based upon the benefits that various social groups receive from environmentally destructive practices and the costs they endure as a result of those same practices (Murphy 1994). Although such a characterization may offer a useful supplement to traditional understandings of class, the vagaries of class analysis in regard to environmental issues suggest that a more fine-grained empirical analysis of the distributional effects of environmental policy is necessary.

Economic Analysis of Distributional Effects

Theorists have come to different conclusions regarding whether blue-collar workers and middle-class environmentalists should be considered allies or adversaries. The contradictory tendencies found in empirical cases suggest that the alliances that emerge in regard to a particular environmental policy cannot be accurately understood in class terms. More detail regarding the specific environmental policy is needed to determine the way in which particular interests are affected in any given case. The general political configuration suggested by class analysis is, however, useful for providing an initial framework. There are three actors of interest here, capital, labor, and the largely middle-class environmental movement. Labor is best understood as situated between capital and environmentalists on issues of policy. Labor is capable of forging alliances with either side, depending on the way in which workers perceive their interests' being affected in a given situation. Such a configuration of capital, labor, and environmentalists was borne out empirically in a national study of labor-environmental relations at the state level (Obach 2002). In this study positive relations between unions and employers were found to be associated with poor relations between labor and the environmental community. This suggests that unions are in the position of choosing sides in regard to environmental issues, and cooperation with one side is related to poorer relations with the other. The question that remains is, what factors influence the choice that unionists will make in a given situation?

Much of the theoretical work on this issue suggests that labor-environmental conflict is more likely than labor-environmental cooperation, given the threat to jobs thought to be posed by environmental protection. Empirical economic analysis, however, suggests a more complicated picture of how the costs and benefits of environmental destruction and protection are distributed. Although some environmental measures may threaten particular jobs, claims of job loss arising from environmental regulations are often politically motivated and unsupported by the evidence (Freudenberg, Wilson, and O'Leary 1998; Goodstein 1999; Kazis and Grossman 1991). Contrary to the allegations of many employers, most research on this issue suggests that on an economy-wide basis, environmental regulation has a *positive* effect on employment rates (Goodstein 1999; Meyer 1992, 1993). According to a report by the Organisation for Economic Cooperation and Development (1997), in economically developed nations "direct and indirect environ-ment-related employment ranges between 1 and 3 percent of the labour force," and although workers in certain industries may be harmed by environmental policies, the net effect of such policies on employment is "on the whole . . . slightly positive" (9).

The positive employment effect of environmental measures can be attributed, in part, to the fact that environmental protection requires the employment of many people. In the United States roughly two million people work in jobs that are directly or indirectly related to environ-mental protection. And although some class theorists suggest that the working class suffers economically as a result of environmental mea-sures, surprisingly, a disproportionate number of the jobs created as a result of environmental policy are blue-collar jobs. In the U.S. economy overall, 20 percent of all jobs are found in the traditional blue-collar sectors of manufacturing and construction. Yet of those employed in jobs related to environmental protection, 31 percent are in construction or manufacturing (Goodstein 1999). Based on an EPA analysis, environ-mental economist Eban Goodstein (1999) reports,

Environmental protection is a largely industrial business. In 1991 . . . the private sector spent $22 billion on pollution control plants and equipment, and $43 billion on pollution control operations, which ranged from sewage and solid waste disposal to the purchase and maintenance of air-pollution-control devices on smokestacks and in vehicles. Federal, state and local governments spent around $12 billion on the construction of municipal sewage facilities. These dollars, not surprisingly, supported a disproportionate number of construction and manufacturing jobs. (37)

Aside from creating the demand for workers to manufacture and build pollution control mechanisms, the employment-generating effect of envi-ronmental policy can manifest itself in a number of other ways. Certain environmental measures, such as pollution control, improve employment prospects by increasing the cost of capital relative to labor, thus increasing the demand for workers and generating employment (Yandle 1985). Jobs are also directly created in environmental restoration (Goodstein 1999; Potier 1986).

The positive association between environmental measures and blue-collar job creation contradicts popular perceptions about widespread labor-environmental conflict. Disputes between environmentalists and development interests, for example, often involve workers in the building trades who support development for the jobs that it creates. Even leaders within these movements recognize the goals of preservation and development as contradictory. According to David Hawkins of the Natural Resources Defense Council, "We've never had good relationships with Building Trades. They see the environmental community as a bunch of people who are going out . . . and stopping development of the land, which is how they make their money. Their members are employed to build things, and they see the environmental community as using all sorts of legal shenanigans to block and slow down projects" (personal communication, June 4, 1999).

Some research suggests, however, that the scope and intensity of development disputes may be exaggerated. A survey of state labor leaders found no evidence to support the belief that the building trades are significantly associated with conflict with the environmental community (Obach 2002). This is consistent with evidence that shows that environmental protection does not necessarily prevent development, it merely affects the way in which development is carried out. As indicated above, environmental measures actually increase the overall level of construction industry employment (Goodstein 1999; OECD 1997). This has at times served as a basis for cooperation between unions and environmentalists. For example, in the state of Washington, environmentalists and the Building Trades worked together to win federal money for habitat restoration, work that would be carried out by workers in the Building Trades. Similarly, in New York the Building Trades worked with environmentalists to ensure that proper environmental precautions were taken in the development and restoration of park land. Although Hawkins indicates that relations between environmentalists and workers in the construction industry are not great, he added that "fights [with the Building Trades] tend to be localized and very sporadic" (personal communication, June 4, 1999). Overall, the mutual interests that environmentalists and building-trades workers have offset the occasional conflicts that occur.

In addition to blue-collar jobs created by environmental protection efforts, employment gains can also be achieved in the service sector. Many lawyers and researchers are employed in environment-related positions. Service-sector employment may increase as the protected natural environment facilitates the growth of the tourist industry (OECD 1997). But despite these positive employment effects, it is questionable whether environmental measures will inspire the active support of service-sector unions. A national study found no consistent relationship between service employment and positive labor-environmental ties (Obach 2002). Although environmental protection has some positive ramifications for service-sector employment, the employment gains that result from such protection may be too distant and uncertain to win the active support of service-sector unions. A tourist industry *may* develop around a protected natural area, and a less polluted community *may* attract white-collar employers seeking to locate in an area with a high quality of life. These outcomes, however, are contingent on numerous other factors besides the enactment of environmental protection measures, and the service-sector employment gains to be had from such measures are far from guaranteed. Potential gains such as these rarely provide enough of an incentive to generate active labor-environmental cooperation.

Another area of employment has been found to have a more direct association with labor support for the environment: public-sector employment. The job-generating effect of environmental protection is directly reflected in public spending. State spending on the environment generates 20–40 percent more employment than the average mix of public expenditures, demonstrating that public employees stand to gain from environmental protection (Meissner 1986). Public employee unions that include workers involved with environmental regulation often maintain cooperative relations with the environmental community (Obach 2002).

Based on these considerations, it is clear that there is no inherent trade-off between jobs and the environment on the economy-wide level or even within most employment sectors. Given that environmental policy tends to benefit both employment and environmental interests, it can serve as a basis for cooperation between workers and environmentalists. Yet the economic analyses discussed above indicate only a *net* job gain from environmental protection and do not address its effects on specific workers. The impact of environmental policy on the economy overall or even on certain employment sectors is less important in determining political conflict than the particular impact of such policy on specific groups of workers. Broad economic improvement or employment expansion is diffused throughout society, and the direct beneficiaries are often unorganized and unspecified prior to the implementation of the employment-generating policy. Potential victims of job loss, however, face concentrated costs and, at least in terms of unionized employees, represent a preexisting politically organized constituency who may oppose environmental measures.

This is the issue that is of most concern when we are attempting to understand labor-environmental relations. According to Goodstein (1999), "The real economy-wide effects of regulation are to shift the *types* of jobs, without increasing the overall level of unemployment" (4, emphasis in original). With the knowledge that environmental regulation has a slightly positive effect on employment *overall*, the question becomes, who benefits and who loses?

In *The Trade-Off Myth*, Goodstein (1999) offers a comprehensive review of the economic effects of environmental regulation. Despite the wild exaggerations found in most industry-sponsored studies and the tendency for all projections to overestimate job loss, his analysis demonstrates that roughly 3,000 jobs are lost annually in the United States because of environmental regulation. This figure represents a tiny fraction of the total number of yearly layoffs: less than one-tenth of 1 percent. It needs to be viewed in perspective relative to other factors that cause mass layoffs. For example, in 1997 almost 80,000 workers lost their jobs as a result of company reorganization, over 10,000 jobs were lost when firms relocated overseas, and 12,000 people were left unemployed as a result of competition from imports. Fewer workers are laid off annually because of environmental regulations than lose their jobs as a result of natural disasters (Goodstein 1999, 46). Recent financial scandals and the subsequent bankruptcies of some major corporations have resulted in more job loss that that caused by environmental protection. For example, over 4,000 Enron employees found themselves out of work when illegal financial irregularities resulted in the collapse of the corporation in 2002.

The minute rate of job loss due to environmental measures may come as a surprise given the disproportionate amount of attention this phenomenon generates. But the fact is that the cost of complying with environmental regulations represents a very small percentage of the overall costs that employers face. Usually this is on the order of 1 to 2 percent, and rarely is the cost over 5 percent. It is very seldom that such costs make enough of a difference in profitability to result in a shutdown. In addition, in many cases the investment required for environmental improvements results in efficiency gains that offset the cost. And as indicated above, any job loss associated with environmental compliance is more than offset by the jobs created by environmental regulation. This can even happen within a single firm, as job losses due to competitive disadvantages resulting from compliance costs are offset by the addition of workers necessary to meet the pollution standards (Goodstein 1999, 53). But of course, such offsetting does not occur in every single case, and unless displaced workers are able to move directly into the new positions created as a result of environmental regulation, this trade-off is of little consolation to them.

Which workers are at risk of losing their jobs as a result of environmental measures? Research from the 1980s indicates that job losses as a result of environmental policies up to that time were concentrated in the metal, chemical, and paper industries (Potier 1986). But as indicated above, actual job loss even in these heavily regulated sectors was minimal. And in those uncommon instances in which facilities have closed as a result of environmental compliance costs, rarely has this been the only reason for the shutdown. Goodstein's analysis indicates that those in the extractive industries, in particular, coal and timber, have really borne the brunt of environmentally induced job loss. Almost 10,000 coal miners were left unemployed in the ten years following the passage of the 1990 Clean Air Act amendments. Timber industry workers lost almost as many jobs when measures were imposed to protect the habitat of the spotted owl in the Pacific Northwest.

Yet even in the highly publicized case of the timber industry, the significance of environmental measures to understanding the plight of the workers is questionable. Despite the hostility directed at the environmental community by displaced workers, a statistical analysis of timber industry employment trends indicates that environmental measures had little to do with job loss in the Pacific Northwest. A study by Freudenberg, Wilson, and O'Leary (1998) demonstrates that timber industry employment had been declining for decades prior to the emergence of the spotted-owl issue. In fact, Freudenberg et al.'s research indicates that, historically, declines in timber industry employment were actually slowed by key environmental legislation, suggesting that rapid depletion of resources, not conservation, is more responsible for job loss in this sector. Several other factors can also be seen as contributing to the plight of timber industry workers during the 1980s and 1990s. In the 1980s increased timber imports from Canada and the drop in the number of housing starts negatively influenced timber industry employment, as did industry strategies such as the export of unprocessed logs, automation, and the shift of timber extraction to the low-wage Southeast (Foster 1993). Thus, even in cases in which environmental protection does contribute to job loss, it is rarely the only or even the most important factor in terms of explaining the plight of the displaced worker. The fact that environmentalists receive a disproportionate share of the blame in these cases raises other issues that will be addressed later.

One might still assume that given the overall job gains that result from environmental policies, workers who do lose their jobs as a result of environmental measures should not face that much difficulty in securing new employment. However, a smooth employment transition is rarely possible for these workers. Oftentimes incongruity between laid-off workers and these newly created jobs in terms of skills and geographic location leaves these workers virtually unemployable. This is especially true for workers in the extractive industries. Several factors make job threats to these workers more intense than those faced by workers in the manufacturing sector. First, coal mining or logging experience generally does not supply a worker with qualifications applicable to other work. The set of skills necessary for this type of labor is very specific compared to that required for most manufacturing jobs, in which skills are more transferable. Second, extractive industries tend to be based in isolated rural communities. Job loss is especially burdensome for workers in such communities, in which few employment alternatives usually exist. Goodstein (1999) also argues that fear of job loss and the associated conflict with environmentalists was more intense in the case of the Pacific Northwest logging industry because the economy of many of the small towns in which the industry was located was based heavily on that one industry. It was thought that the shutdown of this "base industry" would reverberate, causing economic devastation, not just for those workers directly affected, but for entire communities. Larger communities with more diverse economies can absorb job loss in a particular industry, but the economic effect of the shutdown of a single industry is more significant for smaller, more economically homogenous communities.

The intensity of the environmental conflicts surrounding extractive industries is also heightened by the regional concentration of such industries. Such conflicts do on occasion also emerge over manufacturing facilities; several factors, however, tend to limit the intensity of the clash between workers and environmentalists in these cases (Obach 2002). Conflicts over particular polluting plants are typically isolated to the community in which those plants reside. Thus, they have a limited effect on the overall quality of relations between unions and environmentalists beyond those particular communities. Even when national legislation has the potential to affect large numbers of workers in a given industry, if the legislation's impact is distributed over plants scattered around the nation, the conflict between workers and environmental advocates is not as intense as in the case of the regionally concentrated timber industry. Additionally, pollution control strategies are often available to address environmental problems at manufacturing plants while at the same time creating jobs at or near those plants. This creates a realm of cooperation for workers and environmentalists that has no equivalent in the timber industry.

The Pacific Northwest "timber wars" affected an entire region in which many small communities had a concentrated interest in defending existing logging practices. The fact that there were many such communities, all dependent on the same industry, concentrated in one region of the country served to intensify labor-environmental conflict. The strong, tightly knit regional and employment identity among timber workers and their community members served to unite them in the face of what was perceived to be an attack by Washington bureaucrats, countercultural radicals, and elite environmentalists from urban areas (Brown 1995; Dunk 1994). The regional and cultural dimensions of this conflict will be taken up in more detail in a later chapter. But it is clear that the economic impact of environmental measures varies from industry to industry, not only in terms of the likelihood that workers will be affected, but also in terms of how severe the consequences will be. A job lost in the timber industry is more significant than a job lost in manufacturing, because of the manner and extent to which the interests of the workers are affected. This has obvious ramifications for the types of relations that develop between unions and environmental advocates in these two economic sectors.

An interest analysis suggests that the benefit of environmental protection is worth the cost for society as a whole. Research has found that people are willing to pay more as taxpayers and as consumers to cover the costs necessary for environmental protection (Goodstein 1999). Resistance to such protection emerges when the associated costs are disproportionately placed on a relative few. Recognizing that in at least some circumstances workers are forced to bear the bulk of the economic cost of environmental protection, some have advocated programs that would disperse this cost more widely. Job loss always creates hardship for the workers involved, but providing an economic safety net for such workers would reduce this burden. The security provided by such a safety net would alleviate some of the fear threatened workers experience in the face of prospective job loss and might help reduce opposition to environmental measures. This is the logic behind a proposal known as "just transition" (Moberg 1999), whose chief advocate was Tony Mazzocchi, formerly of the Oil, Chemical, and Atomic Workers International Union (OCAW).

Mazzocchi was a leading advocate for workplace safety and the environment, but he recognized the job loss risks associated with measures that increase the cost of production. Refusing to turn away from the need for environmental protection in the face of these risks, he argued that workers who are displaced from their jobs in the name of the environment should be compensated with an extensive retraining program. Such a federally financed retraining program, modeled after the GI Bill designed for workers returning after World War II, would provide displaced workers with a steady income, health benefits, and college tuition money for up to four years after losing their job because of the implementation of environmental policies. This would provide security for such workers and allow them to acquire education and skills to enable them to qualify for high-paying jobs comparable to those that they lost.

Thus far there has been no action to create a retraining program of this magnitude; workers displaced as a result of environmental policies have, however, received compensation in some instances. For example, timber workers who were left unemployed as a result of the expansion of Redwood National Park in 1978 received full salaries for several years after losing their jobs (Mills 1990). In 1988 Congress expanded the Job Training Partnership Act (JTPA), a measure originally designed to aid economically disadvantaged workers, to include assistance programs for workers displaced by environmental measures or other policies. Critics charge, however, that the short-term job training and job search assistance offered through the JTPA does little to benefit workers who have lost good jobs (Goodstein 1999; Oil, Chemical, and Atomic Workers International Union n.d.) and that a much more extensive program is needed to prevent displaced workers from bearing a disproportionate share of the cost of environmental protection. Without such a program, workers in certain sectors will correctly sense that they may be asked to bear the economic burden that results from proposed environmental measures. This perceived threat to their interests is likely to lead to labor-environmental conflict in at least some instances.

Political Interests: A Basis for Cooperation?

As noted in the previous section, the economic interests of workers can be directly affected by environmental measures. But the way in which workers are affected economically by such measures is mediated by the political process through which environmental policies are created. Thus, in examining ways in which worker and environmental interests may coincide, we must also consider political interests and activity. Much coalition theory suggests that workers and environmentalists stand to gain by working together (Bacharach and Lawler 1980; Gamson 1961). In general it is to the advantage of relatively weak organizations to unite with one another to amass the power necessary to overcome adversaries. Organizations that are more powerful have less need to engage in coalition work; thus they tend to refrain from doing so (Hojnacki 1997). Although environmental measures may on occasion threaten the economic interests of certain workers, in general unions and environmentalists share a common powerful adversary, private industry. Given their weakness in relation to their primary political opponent, greater coordination in their respective struggles against employers and polluters would benefit both workers and environmentalists.

Some interest group theory suggests that unions would make particularly effective coalition partners with environmentalists. Unions tend to have more resources to commit to political efforts, but the goals they advance are sometimes viewed as narrow and self-serving. Environmentalists, on the other hand, have fewer resources to dedicate to political action, but they are seen as representing the broader public interest and are thus more capable of garnering public and political support than unions. When unions and environmental groups are able to identify mutually beneficial policies, this combination of resources and a positive public image is thought to be very effective politically (Yandle 1983). As noted earlier, environmentalists and unions have teamed up in certain cases to advance policies that protect the environment through the employment of union workers to perform various cleanup or environmental mitigation projects.

In the electoral sphere, worker and environmental interests also tend to favor coordinated action. When elected officials are rated according to their position on relevant policies, both environmentalists and unions overwhelmingly tend to favor Democratic officeholders. In fact, elected officials who are strongly prolabor also tend to be strongly proenvironment (Goodstein 1999, 14). Activists attempting to build labor-environmental cooperation often cite the common interest that unions and environmentalists have in cooperating in the electoral realm. Yet cooperation in the electoral sphere is hampered by the restrictions placed on nonprofit environmental organizations. Such groups are not permitted to endorse candidates or otherwise engage in partisan politics. This incongruity in terms of political strategy can serve as a barrier to cooperation between unions and environmental SMOs. Yet some environmental organizations forgo their nonprofit tax status in order to engage in electoral politics. Other environmental leaders and activists act independently from their organization in order to participate in this sphere. Unions and environmentalists have coordinated electoral strategy in an effort to promote Democratic candidates. Given their mutual interest in electing Democratic Party candidates, one would expect that Republican political control should foster labor-environmental cooperation, as both the labor and environmental movements seek to oust their common political enemy. However, an analysis of state level intermovement relations indicates that the opposite is true. Labor-environmental relations are actually worse in situations of Republican Party control than when Democrats are in power (Obach 2002). Labor leaders in states where the Republicans dominate politically more often report that they have worse relations with the environmental community than their counterparts in Democrat-controlled states.

One interpretation of this apparently counterintuitive finding is that in the face of adversity, these movement actors seek to secure their own goals and abandon any efforts to work with others to make broader gains. A "unite against the common enemy" approach may be superceded by an "everyone for themselves" strategy. For example, when dealing with a Republican-controlled state government, union leaders may give up on any hopes of securing broadly progressive policies that would satisfy both unionists and the environmental community and instead seek to secure whatever gains are possible for their own constituents. This type of behavior may also be associated with the overall strength of the organizations in each movement. Weak organizations may feel the need to align themselves with those in power and avoid oppositional coalitions in order to secure gains on their own issues. In contrast, when unions and environmentalists are more powerful and able to elect supportive candidates to office, they may then have the luxury of expanding into issues of more peripheral interest to their members. A study of labor-environmental relations at the state level, found no association between union strength and the propensity of unions to engage in cooperative efforts with environmentalists (Obach 2002). Union cooperation with environmentalists is just as common in states in which unions are weak as in those in which they are strong. A more viable interpretation is that in an environment of Democratic Party control, party leaders may help unite unions and environmental advocates to prevent internal conflict and to identify mutually agreeable goals.

Contradictory tendencies emerge when the political behavior of unions and environmentalists is examined in terms of coalition formation. Most coalition theory, however, and many empirical examples indicate that unions and environmentalists stand to gain through greater cooperation. In the policy sphere, they tend to confront the same powerful opposition in the form of employers and polluters. Both unions and environmentalists seek to constrain the practices of the private owners of industry, and the combination of qualities that they bring to the table can make them a significant political force. They also share common interests in the electoral arena with their overwhelming support for Democratic candidates. Their actual behavior in relation to one another under the governance of different political parties requires further analysis, but most evidence suggests that unions and environmentalists would stand to gain through greater coordination in the political realm regardless of which party is in charge.

Conclusion

The relationship between the interests of organized labor and the environmental movement is a very complex one. In looking at broad class interests, some theorists have suggested that the economic interests, of the working class clash with the goals being advanced by the largely middle-class environmental movement. Workers have an interest in economic growth, these theorists point out, and the threat to certain jobs posed by some environmental regulations will drive workers to ally with employers in opposing the environmental movement. Yet others point out that the poor and working class bear a disproportionate share of the cost of environmental degradation in the workplace and in the community. This should drive them into the arms of the environmental movement. In addition, environmentalists who seek to rein in corporate power are natural allies for workers, traditionally locked in struggle with their employers. Overall, on this broad class level, the configuration of interests among workers and environmentalists is contradictory and difficult to gauge.

Economic analysis cuts through some of the vagaries associated with assessments based on broad class interests to suggest that very few workers actually face a threat of job loss due to environmental measures. The overall economic and employment effect of environmental regulation tends to be positive, and workers in some sectors can see immediate concrete gains resulting from environmental measures. This, along with shared concerns over health and safety issues and common political interests, can explain the positive ties that often exist between unions and environmental advocates. Cooperation between these two sectors, although it has received relatively little attention, is far more common than conflict.

But instances of environmentally induced job loss, though rare, can still have a significant impact, not only on the workers directly affected, but also on the general state of relations between unions and environmentalists. Certain employment sectors in which job loss has occurred, such as the timber industry, are associated with poor ties with the environmental community. But the very notion of environmentally induced job loss serves to shape the debate about jobs and the environment, and it can affect the thinking of worker representatives even in sectors in which the possibility of job loss is extremely remote. Employers seeking to avoid costly environmental regulation often appeal to workers on the basis of this job loss fear in order to enlist them in the fight against environmental advocates.

This analysis reveals the general configuration of the relations between environmentalists, workers, and employers. In terms of environmental policy issues, workers can be seen as standing between their employers and environmental advocates. Unions face two basic contradictory pulls, one drawing them to side with the employer in defense of their perceived economic interests, and another drawing them toward a fellow progressive movement whose members are often locked in conflict with the union's own traditional adversary and whose goals are at least purported to be in the common interest. There is a great deal of variability in terms of how these relationships play out. The assertion that the multitude of factors influencing various labor and environmental constituencies can only yield ad hoc relationships has some validity. But under the right circumstances, the potential exists for ad hoc cooperation to grow into a more stable labor-environmental alliance. An assessment of the interests of labor unions and environmental organizations, as presented above, represents only the first step in answering the central questions regarding political conflict or alliance formation between them. An examination of economic and political interests alone will not necessarily tell us about the actual relations between the actors involved. Common interests do not automatically yield cooperation, nor do contradictory goals automatically yield conflict. Thus we must also examine actual cases of conflict and cooperation to see how these interests are acted upon.

In the next chapter the history of labor-environmental relations in the United States will be examined. This will provide a concrete basis on which to assess the extent to which the interests examined in the current chapter influence the actual behavior of the relevant organizations. As will be seen, although basic economic and political factors shape the behavior of these movement actors, relations between unions and environmental organizations are extremely variable and complex. While giving us a basis on which to begin a more thorough analysis of labor-environmental relations, a simple interest assessment cannot provide a full understanding of how these relationships develop. A look at the major national trends in the vacillating relationship between these two movements will offer us a better sense of the variables involved.

3

Labor-Environmental Relations in the United States: A Brief History

Early Alliance Efforts: Health and Energy

Today most political and media attention is focused on instances of conflict between unions and environmental advocates. Yet historically unions can be seen as leaders in the fight for environmental protection (Gottlieb 1993). According to Scott Dewey (1998), prior to the mass popular mobilization around environmental issues in the 1960s, "Labor organizations stood at the forefront of calls for controlling environmental pollution" (47). Initially unions were primarily involved in the environment as a public-health issue. Unions also allied themselves however, at times, with the early preservationist/conservationist environmental movement on issues unrelated to human health. For example, in 1958 the national American Federation of Labor and Congress of Industrial Organizations (AFL-CIO) lined up with the conservation movement in support of the creation of a National Wilderness Preservation System. Still, the majority of early union involvement in the environment was related to public health. This interest in public health issues was manifest primarily in unions' advocacy for air and water protection and their efforts with regard to issues of workplace safety and health. Although some of this support for environmental policies can be tied to employment benefits to be had from certain environmental regulations, there are many instances in which unions backed environmental policies in which they had no specific economic interest. Occasionally unions supported environmental measures even when jobs would be threatened by their implementation (Dewey 1998, 50). Some unions, in particular, the UAW, the Oil, Chemical, and Atomic Workers International Union, and the United Steelworkers, have at times taken a very active stance on environmental measures, occasionally incorporating environmental concerns into collective bargaining and strike demands.

The general tendency for unions to support environmental regulation, or to at least withhold opposition to such regulation, extended through the early 1970s. The Clean Air Act (1970) and the Clean Water Act (1972) both passed with labor support, as did other important pieces of environmental legislation. The UAW was a sponsor of the first Earth Day in 1970, which served as the launching point for several new activist-oriented environmental organizations. Some unions saw promise in the growing environmental movement and sought ways to establish mutual support between this movement and unions. The poor economic conditions that arose in 1973, however, challenged this sentiment toward environmental activism. The early intermittent alignment of labor and environmental concerns was increasingly called into question as economic recession caused unions to focus more intently on job protection and to look with skepticism upon any measure that appeared to threaten this core concern. Calls by some more radical environmentalists for a "zero-growth" economy also contradicted the long-standing labor goal of full employment and rising living standards.

The recession of the 1970s and its negative effect on labor-environmental relations suggests that the general state of the economy is an important factor in shaping union-environmental ties. Prosperous times allow unions and environmental organizations more opportunity to consider values beyond basic economic well-being. But it is still necessary to identify how specific groups of workers are affected and the particular positions adopted by environmental and union actors when the economy is in decline. A closer examination of historical trends reveals that the overall economic condition in the nation is just one factor that influences political action and that more targeted economic impacts and noneconomic issues give rise to a range of responses to environmental measures on the part of labor unions.

By the mid-1970s the now established environmental movement included a host of new organizational actors. The environmental issues raised by these actors pushed well beyond the wilderness protection goals of the earlier conservation movement to include issues such as pollution prevention, toxic-use reduction, and alternative energy. These issues extended the environmental focus into new realms, some of which showed promise for collaboration between environmentalists and unions and some of which created the potential for conflict between the two.

The most obvious area of potential overlap between labor and environmental concerns has always been that of occupational safety and health. It is here that the question of pollution merges with issues of the workplace environment. By the mid-1960s there was a growing awareness of the risks presented by toxic substances both to the environment and to the workers that handled them on the job. Rachel Carson's (1962) groundbreaking book, *Silent Spring*, served as a wake-up call to the American public regarding the dangers associated with chemicals that had become widely used for everything from pest control to manufacturing.

The risk associated with exposure to toxic substances resonated with some labor leaders. The OCAW, under the leadership of legislative director Tony Mazzochi, was at the forefront of making health and safety a central union concern at this time. Several other unions, including the UAW, the United Farm Workers and the United Steelworkers also began to emphasize environmental health and safety issues, owing in part to the hazards they faced in the work environment. Pressure from these unions, along with that exerted by a host of public-advocacy organizations tied to Ralph Nader, led to the passage of the Occupational Safety and Health Act in 1970 and the creation of a federal regulatory mechanism for workplace safety (Gottlieb 1993).

Issues such as toxic contamination were receiving more attention at the time from environmentalists, although most still saw wilderness protection as their primary focus. Several of the large environmental organizations endorsed the passage of the Occupational Safety and Health Act, but there was little active support on the part of these organizations for the legislation, and they generally failed to embrace workplace issues as central environmental concerns (Gordon 1998). The potential for cooperative labor-environmental efforts that this area affords has never been fully realized, yet during this period, some activists tried to build the link between workplace and environmental hazards, and some inroads were made in doing so.

The first major breakthrough in terms of environmental action in support of workers came in 1973 when eleven of the nation's largest environmental organizations offered their support for a strike and boycott by OCAW members against Shell Oil over the issue of workplace health and safety (Gordon 1998; Gottlieb 1993). In the end, the strike did not result in victory for the workers, but it did reveal the potential for alliance building between the labor and environmental movements beyond occasional mutual endorsement of environmental or labor measures. Prior to the Shell strike, support for striking workers on the part of such a broad swath of the environmental movement was unprecedented. After the strike, however, supporters and opponents alike recognized the power of such an alliance, and strategists for a progressive movement attempted to build upon this common ground.

Among the most active organizations involved in these efforts to build bridges between environmental organizations and unions was Environmentalists for Full Employment (EFFE). This organization was founded in 1975 specifically "to publicize the fact that it is possible simultaneously to create jobs, conserve energy and natural resources and protect the environment" (Grossman 1985, 63). EFFE sought to unite the labor and environmental movements and enjoyed some success in bringing unions and environmentalists together both at the grassroots and leadership levels. Labor and environmental activists participated in several joint conferences, and some "letterhead" coalitions included major labor and environmental organizations. But achieving a commitment to active support for mutual goals remained elusive.

Richard Grossman, an EFFE staff member from 1976 until 1984 (the year the organization ceased operations), attributes the lack of substantial success at alliance building to the power of labor leaders who resisted rank-and-file pressure to expand the organizational agenda beyond traditional concerns. Grossman (1985) also argues that in addition to having a narrow focus on work-related issues these leaders fought only defensive workplace battles and constrained militants later on when the Reagan administration launched its attack on unions. A united progressive movement with a common agenda might have proven to be more successful in the years following the 1970s, but the timidity and conservatism of most labor movement leaders prevented such a counteroffensive from being launched. The question of leadership and the strategies it adopts are certainly relevant to the question of coalition formation. Later changes in labor leadership indicate the importance of the direction set by those at the top of the organizational hierarchy. This topic will be addressed in more detail in a later chapter, but it is important to recognize that although an organization's leadership can act as a conservative force, it is also possible for leaders to advance organizational change. Leaders are important actors in coalition politics, and the direction in which their influence will exert itself cannot be accurately generalized. Yet in the 1970s leadership within the labor movement was largely conservative. The economic conditions at the time also fostered a reluctance on the part of these leaders to branch out into issues with uncertain economic implications.

EFFE's choice of issues made its work particularly challenging during the period of its existence. EFFE placed a great deal of emphasis on safe energy, which proved to be a difficult issue on which to forge intermovement cooperation. It called for a decentralized renewable energy system and campaigned against nuclear power, a popular issue among environmentalists, but one that never fully caught on in the labor movement. Despite the dangers nuclear facilities posed to workers and communities, the energy crisis of the early 1970s and a slow economy led most union leaders to view nuclear energy as necessary for economic growth and the jobs it would bring. The building trades, whose members build nuclear facilities, exerted much of the pronuclear pressure, but the rest of the labor movement did little to challenge that position. To the disappointment of the environmental community, the AFL-CIO, after a 1976 conference on energy, drew the conclusion that the "rapid development of nuclear power is a 'must' without which the nation's economy would falter" (*AFL-CIO News*, March 27, 1976, 1, as quoted in Logan and Nelkin 1980, 6). With few exceptions the labor movement remained decidedly pronuclear through the 1970s, and broad labor-environmental alliance building was rejected in favor of ad hoc cooperation on isolated issues of mutual concern (Logan and Nelkin 1980).

Although activists during the 1970s were generally unsuccessful at building a permanent alliance between unions and environmental advocates, workplace safety issues did allow for some cooperation between the two. For example, environmentalists offered support for a United Mine Workers strike in 1979 over issues of health and safety (Grossman 1985). EFFE and a parallel organization, the Labor Committee for Safe Energy and Full Employment, were eventually successful at building some labor-environmental ties around the hazards of nuclear energy, including a march of 15,000 union members and environmentalists on the second anniversary of the Three Mile Island nuclear accident. But cooperative actions on such issues remained intermittent throughout the 1970s while union concerns over the economy and jobs continued to overshadow "tangential" issues such as the environment.

Reagan, Regulation, and the Right to Know

Some critics charge that during the 1980s the mainstream environmental movement adopted an increasingly conservative and narrow single-issue focus (Dowie 1995; Dreiling 1998). Unions for the most part also adopted a conservative, defensive posture during this period as corporate attacks and a decline in manufacturing decimated their ranks. Retaining jobs remained their central concern. Although the tendency toward a limited issue focus on the part of social movement organizations can be seen as a product of the U.S. legal and political structure, as discussed further in chapter 5, shared concerns and cooperative efforts between unions and environmentalists were not nonexistent during this period. The political developments that fostered defensiveness on the part of both unions and environmentalists also opened up grounds for cooperation during this period.

Although some unions were slow to embrace workplace health and safety as a central concern, labor leaders and rank-and-file activists had come to recognize the value of the Occupational Safety and Health Administration (OSHA) and the federal workplace standards that had been put in place. At the same time environmentalists were becoming adept at using the EPA to advance their goals. Upon the election of the militantly pro–free-market Ronald Reagan to the presidency in 1980, mutual concern about the new administration's commitment to the government's regulatory responsibilities reignited efforts to forge labor-environmental cooperation. Under the direction of the AFL-CIO's Industrial Union Department and some national environmental organizations such as the Sierra Club, unions and environmentalists pledged to work together in Washington while forming statewide labor-environmental networks to defend against rollbacks in workplace and environmental regulation. These statewide networks were instrumental in fighting against deregulation at the national level. But they were also designed to press for expanded regulation at the state level with the knowledge that a patchwork of state regulations would lead frustrated industry leaders to drop opposition to comprehensive national legislation.

The greatest strides were made in the area of right-to-know laws (Obach 1999). Right-to-know measures require that workers and community members be informed about hazardous substances being used in production processes. Such knowledge enables workers to protect themselves against workplace hazards, while also allowing environmental activists to better guard against contamination in the community. Campaigns to have right-to-know laws enacted reflected the environmental movement's growing focus on toxic substances, which increasingly proved to be fertile ground for cooperation with unions. The Love Canal toxic-waste disaster in 1978 inspired some unions not only to dedicate energy to workplace environmental issues, but also to give more serious consideration to the effects of toxic pollution on the communities in which union members and their families lived.

State-based labor-environmental networks were primarily designed to protect regulatory mechanisms and to advance the cause of environmental health both in the workplace and in the community. In some instances, however, the cooperative efforts of these networks expanded beyond their initial purpose. In Wisconsin, for example, the Labor-Environmental Network went on to address issues such as recycling and trade (Obach 1999). That coalition continued to function into the 1990s, and relations between the labor and environmental communities in that state remain positive.

Job loss fears during the 1980s prevented unions from fully embracing environmental concerns, and the mainstream environmental movement adopted some conservative and defensive tendencies during this period as well. Yet common support for workplace and environmental regulation did allow for some mutual effort between unions and environmental groups during this time. Safety and health issues remain the most obvious area of interest overlap between the labor and environmental communities today (Obach 2002). Cooperation between movement organizations is likely to emerge when such overlap exists, regardless of how narrow the focus of the respective organizations. But cases of instrumental cooperation between organizations can give rise to a broader understanding of social problems among coalition partners, thus allowing for greater cooperation. Both of these themes will be revisited in later chapters.

The Fight in the Forest

While regulatory issues and safety and health concerns were providing common ground, during the 1980s an issue was brewing in the Pacific Northwest that would prove to be broadly detrimental to labor-environmental relations. This was a classic jobs-versus-the-environment dispute involving the timber industry in which thousands of timber industry jobs were alleged to be threatened by environmental measures designed to protect the few remaining old-growth forests in the United States. The dispute lent itself to sensationalist coverage because the media were able to simplify the battle as one over a single threatened species, the northern spotted owl (Cooper 1992; Fitzgerald 1992; Gup 1990). The creative antics of some radical environmentalists in the region also generated attractive footage and served as the basis for stereotypic images of countercultural environmental activists battling blue-collar workers. More than any other, this dispute served to shape public perceptions regarding the trade-off between jobs and environmental protection, in addition to leaving a lasting impression on labor leaders throughout the nation regarding the threat to their interests posed by environmentalists (Obach 2002).

Although the conflict over timber harvesting in old-growth forests reached a peak in the late 1980s and early 1990s, battles between workers and environmental advocates over forest issues date back to the 1970s, when Congress sought to expand Redwood National Park in northern California. This effort to expand the park incited timber industry workers to drive logging trucks around San Francisco's Federal Office

Building in protest (Miller 1980). But the major timber industry conflicts took place during the 1980s and 90s around the protection of old growth forests. The economic stagnation of the early 1980s, which included a major slump in housing starts, along with the general antiregulatory orientation of the Reagan administration, led to policies designed to increase cutting dramatically in the remaining stands of old-growth forest on federal land. This increased cutting met with strong opposition from environmentalists, who adopted tactics ranging from tree spiking to federal lawsuits.

While radical environmentalists associated with the direct-action group Earth First! were engaging in tactics such as blocking logging trucks and sabotaging logging equipment in response to the increased harvesting, mainstream environmental organizations, such as the Sierra Club Legal Defense Fund, filed lawsuits and administrative appeals to halt the accelerated cutting. In a demonstration of the serious impact that old-growth cutting would have on the forest ecosystem, they were eventually successful in having a rare forest-dwelling owl, the northern spotted owl, designated as a threatened species in 1990. Environmentalists were able to establish that the survival of this owl would be imperiled if the cutting of old-growth forests continued at the accelerated rate. The official listing of the spotted owl required that the federal government follow a set of procedures delineated in the Endangered Species Act, including analysis of the threat to the owl and the development of a plan to protect its habitat, old-growth forests. Government and outside analyses indicated that millions of acres of federal timberland would have to be set aside for preservation in order to protect enough of the remaining spotted owls to sustain the species. Whereas the implementation of such a plan would represent a tremendous victory for environmentalists, estimates of timber industry job loss based on the various preservation proposals ranged from 15,000 to over 100,000 (Foster 1993).

Although the primary conflict of interests in this matter was between those of the timber industry and those of environmental advocates, it was the federal government (the Bush administration in particular) and environmentalists who engaged in direct legal struggle. Yet to many workers in the logging industry, it was a struggle between jobs and the environment, workers versus environmentalists. In their eyes the need to protect the spotted owl would "lock up" much of the remaining forest land, resulting in widespread job loss. Despite the fact that several other factors contributed to job loss in the region (see chapter 2), hostility toward environmentalists ran high in timber-dependent communities throughout the Pacific Northwest.

In his analysis of this struggle, John Bellamy Foster (1993) attributes the division between environmentalists and timber workers to a lack of class considerations on the part of the largely middle-class environmental movement. He charges that environmentalists failed to support workers during labor disputes with industry in the mid-1980s, a move that would have built trust and facilitated labor-environmental cooperation around the spotted-owl issue. Environmentalists also failed, according to Foster, to incorporate any serious consideration for job loss into their demands for forest protection. This made environmentalists appear insensitive to worker concerns and enabled timber industry leaders and their Bush administration allies to depict environmentalists as the primary enemy of the worker. The industry made strategic use of this opportunity by holding antipreservationist training sessions for workers during work hours at mills and by supporting worker actions against the environmental cause. There is even evidence to suggest that a policy proposal that would have protected spotted-owl habitat while limiting job loss was suppressed by the Bush administration to perpetuate the labor-environmental division (Foster 1993).

Ultimately, the economic devastation of the timber region of the Pacific Northwest as a result of measures to protect old-growth forests did not play out as predicted by the critics of such measures. Despite efforts by the Bush administration to use this issue as a wedge to draw blue-collar support for his reelection campaign in 1992, George Bush was defeated in his bid for reelection by Bill Clinton, who offered a more moderate approach to the issue. President Clinton implemented a plan in 1993 that called for restrictions on cutting in federal old-growth forests along with an economic-assistance package for the affected workers and their communities. In some ways this program went beyond the JTPA, the relatively weak measure already in place to assist displaced workers. Clinton's plan for enhanced job retraining assistance included economic aid to the affected communities and an innovative Jobs in the Woods program designed to retrain timber industry workers to be skilled ecosystem managers who could command high wages while conducting environmental restoration.

Although this program enjoyed only a modicum of success, the Pacific Northwest underwent an economic boom in the 1990s which did more for laid-off timber industry workers than the government programs. Consistent with the evidence regarding the economic benefits of environmentalism, forest preservation measures may have even played a role in this economic expansion. Businesses were attracted to the Pacific Northwest by the quality of life offered in that part of the country, including the pristine environment. As a result, most unemployed timber industry workers were able to find new employment, although many experienced wage cuts as their new jobs did not offer the same level of compensation as their previous ones.

In all, roughly 10,000 workers in the region lost their timber industry jobs in association with the preservation measures undertaken to protect old-growth forests in the Pacific Northwest. Even though these losses were offset in other areas and the displaced workers found new jobs relatively easily, the spotted-owl controversy, more than any other issue, has shaped the perception that blue-collar workers are made to pay the price for environmental protection.

Coal Miners and Clean Air

Another case from the same time period also generated conflict between unions and environmental advocates. In this case the issue was the revision of the Clear Air Act, the main opponent of which was the United Mine Workers. Prior to 1970 there was little federal action on air quality. A few relatively weak measures regulating air quality had passed in the 1950s and 1960s, but the Clean Air Act of 1970 was the first significant piece of federal air quality legislation. This legislation, passed when new support for environmental protection was strong, had the backing of organized labor and, other than some objections from automakers, met with surprisingly little opposition from industry (Bryner 1995). The central feature of this legislation was the development of national air quality standards and requirements that states develop plans for meeting them. At the time, these provisions of the act were relatively noncontroversial, and the positive effects of this legislation on air quality and human health are broadly viewed as a success. According to a 1999 analysis by the EPA, the economic benefits of the Clean Air Act in terms of human and environmental health has exceeded the costs by a ratio of four to one.

When the act was amended in 1977, labor split with environmentalists over automobile emission standards, but they made common cause on most elements of the legislation (Morriss 2000). The factors behind this concurrence demonstrate the complexity of the interests involved in environmental issues and the failure of the simplistic jobs-environment trade-off perspective. Labor and environmental interests converged in two ways with regard to the 1977 amendments, both having to do with the regional impact of certain provisions of the law and the strength of organized labor in different parts of the country. First, the amendments included a measure that held new sources of pollution to higher standards than existing facilities. Since all localities had to meet the same air quality standards under the act, one risk faced by the unionized workers of the Midwest and Northeast was that the less industrially developed parts of the country, such as the South, would have a comparative edge in attracting new investment if polluters were allowed to take advantage of the comparatively unpolluted air in those regions. Without the more stringent requirements for new pollution sources, new facilities in an already polluted northern city would have to limit their emissions a great deal in order for the area to remain in compliance with the act, whereas those built in nonpolluted areas would have more latitude in respect to emissions. This would favor investment in the South and West, where unions were weak and where state laws were unfavorable to unionization. By requiring that new facilities meet higher emissions standards, the legislation effectively eliminated industry's incentive to abandon the already polluted unionized industrial areas.

Second, the 1977 amendments also included a provision that would unite environmentalists with their future adversaries, coal miners. One pollutant that regulators were attempting to limit was sulfur dioxide. The coal mined in eastern states such as Ohio, Kentucky, and West Virginia contained relatively high levels of sulfur that, when the coal was burned, released sulfur dioxide into the air. There were two possible means by which sulfur dioxide emissions could be reduced. The first, a technological solution, was to install scrubbers in the smokestacks of coal-burning plants that would capture the sulfur dioxide before it was released into the air. The second option was to switch the coal being used to that with a lower sulfur content, such as that produced by the primarily nonunion mines in the West. The second option would have harmed coal miners in the eastern stronghold of the United Mine Workers. The 1977 amendments incorporated the technological solution, requiring that all coal-burning facilities install scrubbers to reduce sulfur dioxide emissions regardless of the sulfur content of the coal being burned. This again eliminated the comparative advantage of nonunion western coal, and environmentalists and the United Mine Workers rallied behind the legislation including the scrubber requirement (Ackerman and Hassler 1981).

Yet the unity that was achieved in 1977 was not repeated in 1990 when the Clean Air Act was again amended. The 1990 legislation dealt with a wide range of issues, from toxic emissions to ozone depletion to acid rain. It enjoyed the support of some labor unions, including the United Steelworkers, who joined environmentalists in the National Clean Air Coalition, but the United Mine Workers strongly opposed the legislation's acid-rain provisions. Research during the 1980s had focused greater attention on the problem of acid rain. For years scientists had identified declines in fish stocks in mountain lakes, some of which were declared dead and unable to support native fish life. There was a corresponding decline in the health of many forests. These problems were attributed to the increasing acidity of water and soil in the areas, which was ultimately linked to the continued release of sulfur dioxide into the air. Although the damage from acid rain was occurring primarily in the Northeast and in parts of Canada, the culprit was found to be coal-burning energy and manufacturing facilities in the Midwest. Once in the atmosphere, the sulfur dioxide released from coal-burning plants in the Midwest is carried eastward by the wind for hundreds of miles and returns to the earth via the rain. It is this acidic rain that has been found to have a detrimental effect on lake and forest ecosystems.

Additional changes to the Clean Air Act had been blocked for years since the 1977 amendments. Senate Majority Leader Robert Byrd of West

Virginia had used his position to stifle many attempts to expand the Clean Air Act on the grounds that sulfur restrictions would result in the loss of coal-mining jobs in his state. Like timber industry workers in the Pacific Northwest, eastern coal miners had already faced tremendous job loss. Between 1978 and 1990 productivity in coal mining had more than doubled through technological innovation, greatly reducing the demand for labor. In addition the more labor-intensive underground mining carried out in the East was being displaced by far less labor-intensive strip mining, a practice more common in the West. Once again, job insecurity unrelated to environmental protection was exploited and redirected by industry to serve its own antienvironmental interests. Industry-sponsored research predicted that a minimum of 200,000 jobs would ultimately be lost as a result of the proposed amendments (Goodstein 1999).

Despite these dire predictions, a change in the Senate leadership and a need for the self-proclaimed "environmental president," George Bush, to enact some major environmental legislation led to movement on amending the Clean Air Act. Opposition from the coal industry and the United Mine Workers was not enough to prevent sulfur restrictions from being incorporated into the 1990 measure. This time, however, the restrictions would not include the technological fix around which workers and environmentalists had united in the past. Consistent with the policy trend in the Reagan and Bush administrations in favor of flexibility and market-based regulatory mechanisms, a system of tradable pollution permits was created in order to reduce sulfur dioxide emissions. To reduce overall emissions, firms could lower their own levels of sulfur emission or purchase tradable pollution permits issued in limited numbers by the government in accordance with clean-air goals. In terms of reducing their own sulfur dioxide emissions, coal-fired plants were granted the flexibility to switch to low-sulfur coal if it proved to be more economically feasible than the use of smokestack scrubbers. As a result many firms switched to low-sulfur coal, and thousands of eastern coal miners lost their jobs.

The number of workers affected by job losses resulting from the 1990 amendments was not nearly as high as industry studies had predicted, however. Between 1990 and 1997, approximately 7,000 coal miners were put out of work, most of them in the East. As is typical of environmental regulation, the result was a redistribution of jobs, not an absolute loss. Coal mining in the West and rail shipping boomed in the aftermath of the Clean Air Act amendments (Goodstein 1999, 43). Nonetheless, the United Mine Workers lost thousands of members, and environmentalists were blamed for these losses. A slightly expanded job training program designed for displaced miners was included in the 1990 Clean Air package. However, unlike timber workers, who partially recovered during the Northwest's economic boom of the 1990s, short-term training and job assistance did little to help unemployed coal miners living in small coal towns in the East.

Conflicts between environmentalists and coal and timber industry workers in the late 1980s and early 1990s cast a pall over labor-environmental relations in general. These conflicts firmly entrenched the notion that environmental protection required a trade-off in jobs, one in which blue-collar workers would suffer. Economic analysis indicates that the relationship between jobs and the environment is not that simple, that relatively few workers actually lose their jobs as a result of the implementation of environmental measures, and that more workers benefit from such measures than are harmed by them. Even in the timber and coal cases, the total number of jobs lost was relatively small and did not approach the predictions offered by critics of environmental action. Yet these issues commanded enough attention to shape the perceptions of the public and even those of labor and environmental leaders. The dominant view was that the relationship between the labor and environmental sectors was largely adversarial. Issues would emerge in the mid-1990s, however, that would once again bring the two movements together.

Trade, Teamsters, and Turtles

In terms of labor-environmental relations, the issues that received the most attention in the mid- to late 1990s have been those associated with international trade. During the 1990s several forces converged to make the expansion of international trade a top priority of many economic and political elites. Throughout the 1980s the U.S. trade deficit grew dramatically. Policymakers sought to reduce this deficit by eliminating what they considered to be barriers to the sale of U.S.-made goods abroad. The collapse of the Soviet Union also contributed to the advance of free trade and neoliberal policies in general. The loss of the alternative model of economic organization offered by the Soviet Union empowered neoliberal ideologues to seek to advance unrestricted capitalist relations globally. In addition, the protectionist policies that had been tolerated in America's less developed Cold War allies could now be dismantled in favor of free trade and a full-scale economic restructuring in accordance with neoliberal thinking. With the convergence of these factors the expansion of global trade became the top priority on the agenda of multinational corporations. The opening up to trade and the general transformation of the national economies of former socialist states and the less-developed nations was promoted primarily through international financial bodies such as the World Bank and the International Monetary Fund (IMF). Although these institutions had begun advocating trade liberalization in the third world as early as the 1980s, liberalization policies became even more widespread and more forcefully imposed in the 1990s.

In addition to working through the IMF and World Bank, in the 1980s the United States had begun using retaliatory sanctions against nations whose markets were viewed as not sufficiently open to U.S. goods. At the time these bilateral actions did not incite the opposition of labor or environmental leaders. It was the regional- and global-trade agreements that drew the most attention from unions and environmentalists in the United States. In particular, the North American Free Trade Agreement (NAFTA) and the World Trade Organization (WTO) have met with a great deal of resistance from both sectors.

The United States entered into a free-trade agreement with Canada in 1988. This agreement generated relatively little controversy in the United States; when the proposal was made to extend the agreement to Mexico, however, significant resistance emerged. Both unions and some segments of the environmental community identified problems with the NAFTA proposal. They feared that state and national labor and environmental standards would be undermined by the free flow of capital between the United States and Mexico. They argued that investment would tend to gravitate toward nations with weak standards and low wages. In the words of independent presidential candidate and NAFTA opponent Ross Perot, a "giant sucking sound" would be heard as jobs and investment were swept into Mexico. U.S. and Canadian firms would seek to avoid costly regulations and high wage rates by relocating to Mexico, and those that did not move would be driven out of business in the face of competition from low-wage Mexican producers. This would in turn create pressure on the United States and Canada to curtail regulations in order to accommodate industry, thus setting off a "race to the bottom" in which labor and environmental standards would deteriorate across the three nations (Brecher and Costello 1994; Schaeffer 1997).

In addition to the pressure to "harmonize downward" in terms of wages and standards, NAFTA included provisions that allowed member states to challenge "nontariff trade barriers," national or state regulations that in some way inhibited profit making by foreign producers exporting goods into the nation. Environmentalists feared that in some cases environmental measures could be challenged if they were found to infringe upon the ability of a manufacturer in another member country to sell their goods. This represented a direct threat to the standards that environmentalists were working to institute.

Although unions and environmentalists had distinct concerns in regard to NAFTA, the common threat the agreement presented created the impetus for labor-environmental cooperation. This took the form of two national coalitions that included many of the major labor and environmental actors, the Citizens Trade Campaign and the Alliance for Responsible Trade. The positioning of unions and environmental advocates on this issue and within these coalitions is complex. Some labor organizations and environmental groups were active participants in the coalitions; others worked independently to oppose NAFTA; and some environmental organizations, although seeking modifications in the terms of the agreement, in the end endorsed it.

Michael Dreiling and Ian Robinson (1998) examined the mobilization of different unions against NAFTA and their willingness to enter into coalitions with environmental organizations. They do so on the basis of two strategic or ideological characteristics, a union's general "inclusivity," meaning the extent to which it is oriented toward broad class goals, as opposed to focusing on narrow member interests, and its "radical-ness," the extent to which the union's leaders accept existing political and economic arrangements. Dreiling and Robinson found that unions with a "social union" (inclusive but not radical) or "social movement union" (inclusive and radical) orientation were more inclined to be active participants in the anti-NAFTA fight and more willing to work with their environmental counterparts compared to those with a "business union" orientation (neither inclusive nor radical). Others have also identified the importance of unions' ideological or strategic orientations in attempts to understand labor participation in coalition activity (Dreiling 1998; Dreiling and Robertson 1998; Seidman 1994; Siegmann 1985).

At times unions' ideological or strategic orientations have been found to be more important in understanding coalition propensities than the actual stake that workers in different sectors have in a particular policy. This contrast can be seen in looking at the involvement of the Service Employees International Union (SEIU) and the building-trades unions in relation to the NAFTA struggle. Neither service workers nor building-trades workers faced a significant threat from NAFTA. They both represent workers whose jobs are unlikely to be directly affected by expanded trade. Most of the building-trades unions had little involve-ment in the anti-NAFTA struggle. This is consistent with the business union orientation that characterizes the building-trades segment of the labor movement: Where their own members' interests are unaffected, they tend to be uninvolved. The SEIU, in contrast, was active in the anti-NAFTA campaign, despite NAFTA's lack of direct impact on their members. This reflects the social unionism embraced by the SEIU and its leadership. SEIU's involvement in the struggle against NAFTA was undertaken in the name of solidarity with other working-class people who were threatened by the trade agreement.

Of course, attributing the behavior of these unions to ideology only raises the question of why different unions have embraced contrasting ideologies. A more extensive historical analysis would be necessary to fully explore this question, but there are some obvious factors to con-sider in this regard. For example, the business union orientation found among the building trades can be tied to their historical roots in the American Federation of Labor (AFL), an early federation composed of skilled craft union workers who only later merged with the more radical unskilled industrial workers union federation, the Congress of Industrial Organizations (CIO). The AFL was organized along craft lines and used a strategy in which members were able to advance their interests by monopolizing their unique skills. This focus on serving the interests of their particular members stands in contrast to the efforts of the industrial unions that organized unskilled workers. Unable to rely upon a strategy of monopolizing scarce skills, the CIO adopted a broader class based approach in which they sought to organize the entire mass of unskilled workers. These historical distinctions explain some of the differences between the industrial and craft unions still seen today.

The behavior of the environmental community demonstrated still greater complexity in relation to the free-trade issue, although some suggest that ideological factors are just as important in understanding the response of environmental organizations to NAFTA. The NAFTA debate can be seen as reflective of an issue that has long challenged the environmental community, the issue of economic growth. John Audley (1995) suggests that environmental organizations demonstrate a range of views regarding economic growth, with some equating growth with ecological destruction and others taking the view that growth and environmental protection are not mutually exclusive. Given the economic growth expected to result from liberal trade policies, environmental organizations' views on NAFTA can be understood accordingly, with growth skeptics taking an adversarial position toward NAFTA and others seeking accommodation.

Several of the major national environmental organizations that express the need for economic growth, such as the National Wildlife Federation, the Environmental Defense Fund, and the World Wildlife Fund, came out in support of NAFTA. Because they did not reject the trade treaty outright, they were able to maintain ties with policy elites involved in crafting the agreement. They used their insider status to try to incorporate environmental protections into the trade pact, and although they were successful in including some environmental measures in a "side agreement," they essentially failed to win the inclusion of enforceable environmental policies in NAFTA. Others in the environmental community, such as the Sierra Club and Friends of the Earth, took an adversarial position on NAFTA. They aligned themselves with unions and other opponents in an effort to defeat the trade agreement outright.

Michael Dreiling (1997) offers an alternative interpretation of the positioning of environmentalists on the NAFTA issue. He suggests that the growth of the environmental-justice movement, with its focus on the issues of race and class and the distributional dimension of environmental hazards, had successfully pushed some mainstream environmental organizations to incorporate more social concerns into their work. Some national environmental groups resisted the types of reform advocated by the environmental-justice movement and maintained their traditional focus on land use issues combined with a growing embrace of market-based policy solutions. These and other organizations formed the Environmental Coalition for NAFTA and pressed for the passage of the accord, along with the side agreements they had negotiated. But others responded differently to the environmental-justice appeal. According to Dreiling:

The challenge of environmental justice has introduced not only a political ideology that is critical of the environmental mainstream, but has urged that movement practices be aimed away from conventional channels of power towards the grassroots. Emulating some of these environmental justice tendencies during the NAFTA fight, groups such as Greenpeace and Friends of the Earth (FOE) pursued a strategy of international solidarity and cross-movement alliance building. (69)

Thus these environmentalists joined with unions and others to oppose NAFTA, not only because of its potentially devastating environmental consequences, but also because of its harmful social ramifications. Although the NAFTA struggle created some rifts within the environmental movement (Dowie 1995), it also demonstrated to labor that at least some elements in the environmental community could be worthy allies. Despite the fact that participation in coalitions around this issue was not universal, labor-environmental alliances against NAFTA indicated an important new realm of issue overlap that would prove to strengthen ties between some key labor and environmental actors.

NAFTA was just the first of several instances of trade-related labor-environmental alignment. Unions and environmentalists also worked together to oppose fast-track trade authority for the Clinton administration. The fast-track provision allows the president to negotiate trade agreements that are then presented to Congress as a package that can be either adopted or rejected but cannot be amended. This was designed to prevent negotiated agreements between nations from being subverted by provisions added on by members of Congress during the ratification stage. Allowing only a simple up or down vote tends to favor the passage of trade agreements because reservations about jobs or the environment are typically not great enough to override the generally protrade sentiment found in Congress. Although fast-track authority was granted to President Bush for negotiating NAFTA in 1991, united labor and environmental opposition to a hemispheric trade agreement thwarted fast-track authority in 1998 (MacArthur 2000).

The most dramatic example seen so far of labor-environmental cooperation came about as a result of another trade-related issue. This came at the meeting of the WTO in Seattle in 1999. The WTO, the successor organization to the General Agreement on Tariffs and Trade (GATT), is designed to establish rules governing trade and reduce trade barriers among more than one hundred member nations. In addition to regional trade agreements, through the GATT and then the WTO the US sought to expand international trade further. Fearing a race to the bottom on a global scale, unions and environmentalists mobilized to oppose the advance of the neoliberal agenda at the WTO meeting in Seattle. Numerous environmental organizations and unions coordinated their efforts for a massive demonstration at the WTO meetings. Labor and environmental mobilization for this demonstration was coordinated in part through a new coalition called the Alliance for Sustainable Jobs and the Environment (ASJE) (Carlton 1999). The coalition included some national environmental groups such as the Sierra Club, Friends of the Earth, and Earth Island Institute in addition to several unions, including representatives from the United Steelworkers, the Teamsters, and unions representing postal workers and Teachers.

The United Steelworkers and environmentalists in the Pacific Northwest were central actors in the ASJE. Cooperation between the two grew out of their common conflict with the Maxxam Corporation and its CEO, Charles Hurwitz. Maxxam, which owns Pacific Lumber, had been the target of environmentalists because of its involvement in the cutting of old-growth forests. Maxxam is also the parent corporation of Kaiser

Aluminum, a firm with which the United Steelworkers were engaged in a bitter labor dispute. These struggles were further linked, beyond the common ownership of the targeted corporations, when some striking steelworkers were replaced by laid-off Pacific Lumber employees, highlighting Maxxam in stark terms as the common enemy of unions and environmentalists.

The cooperative effort between environmentalists and the United Steelworkers in fighting Maxxam was expanded into the alliance, which then took on other issues of common interest. Participants in the alliance identified corporate power as the source of exploitation of both workers and the environment. Strong government regulation was viewed as the means by which that abuse could best be curtailed. According to the "Houston Principles," the founding document of the ASJE:

Corporations have become more powerful than the government entities designed to regulate them. . . . Too often, corporate leaders regard working people, communities, and the natural world as resources to be used and thrown away. . . . Recognizing the tremendous stakes, labor unions and environmental advocates are beginning to recognize our common ground. . . . While we may not agree on everything, we are determined to accelerate our efforts to make alliances as often as possible. (Alliance for Sustainable Jobs and the Environment 1999, 2)

Based on the analysis put forth by the alliance, trade liberalization and the WTO were prime targets of alliance action because of the deleterious effect that they have on the ability of individual governments to regulate corporate abuses. In another example of the effectiveness of labor-environmental cooperation, the alliance's efforts in Seattle were largely successful, at least in the short term. The more than 30,000 demonstrators who turned out at the event effectively prevented any advancement in the global-trade negotiations at the Seattle meetings and slowed progress on those negotiations subsequently. Although violent clashes between a relative handful of unruly demonstrators and riot police grabbed headlines, the significance of this mobilized labor-environmental coalition was recognized by many. Headlines heralded the new alliance of "Teamsters and Turtles," referring to the union truck drivers and environmentalists dressed as endangered sea turtles who marched together in Seattle.

These alliances between environmental organizations and unions over the free-trade issue reflect instances in which the fairly distinct concerns of unions and environmental advocates are both negatively affected by a single policy. Union concern for jobs and labor standards as a result of trade liberalization were matched by environmental concern for the downward leveling of environmental regulations that seemed likely to result from such liberalization. Both unions and environmental organizations saw the threat posed by unregulated trade, and they joined forces to resist it. As with the alliances that emerged to oppose Reagan's anti-regulatory agenda, this sort of instrumental cooperation is the most common basis for political coalitions. Yet these cases indicate that labor-environmental cooperation is more complex than a simple alignment of instrumental goals. The specific approach of different environmental organizations and the ideologies embraced by each union also play a role in determining whether or not alliances will form between the two. It is also clear that experience combating a common enemy, as in the case of the environmental-steelworker alliance against Maxxam, can generate a deeper analysis and cooperation in other areas beyond the initial issue that brought them together (Alliance for Sustainable Jobs and the Environment 1999, 2000). These issues emerge as important themes addressed in later chapters.

Continuing Challenges: CAFE and ANWR

Although trade served as the primary issue uniting unions and environmentalists during the 1990s, other issues continued to crop up that undermined this cooperation. One source of conflict between environmental and labor organizations focuses on proposals for strengthening federal CAFE standards. These air pollution standards, initially created in 1975 as part of the Energy Policy and Conservation Act, require that the car fleets produced by U.S. auto manufacturers meet a minimum average standard for fuel economy across the entire line of vehicles a particular manufacturer produces. The standard of 27.5 miles per gallon (mpg) for cars and 20.7 mpg for light trucks has remained unchanged since 1985.

The CAFE standards effectively improved the fuel efficiency of American car fleets during their first decade, but changes in the auto market since the mid-1980s have resulted in backsliding. Despite dramatic improvements in fuel economy technology, since 1987 overall fuel effi-ciency for new vehicles sold in the United States has actually declined by 7 percent. This reversal of previous improvements is largely due to a dra-matic increase in the sale of sport utility vehicles (SUVs) in place of the more fuel-efficient traditional passenger cars. SUVs are classified as light trucks; thus they are required to meet the lower CAFE standard for those vehicles. The disproportionate rise in the number of SUVs sold relative to cars has resulted in the overall decline in fuel efficiency (Friedman et al. 2001). To correct for this unexpected development several envi-ronmental organizations have made strengthening of the CAFE stan-dards a priority. They seek to eliminate the separate categories for cars and light trucks, creating a single combined average for fuel efficiency that must be met by each firm.

In 2001, environmentalists were successful at getting a measure aimed at creating a single standard introduced in Congress. Democratic Sena-tors John Kerry and Ernest "Fritz" Hollings advocated raising the CAFE standards for the combined fleet of cars and light trucks to 35 miles per gallon by 2013. This met with great resistance from U.S. automakers, who derive a significant portion of their profit on their popular SUVs. They feared that a combined standard for cars and light trucks would improve the competitive position of passenger cars, thus cutting into the SUV market. Manufacturers made dramatic claims about job loss if the proposed changes in fuel economy standards were implemented (Sierra Club 2002). Once again industry leaders sought the union to serve as a sympathetic coalition ally in their fight against the new standards. Automakers enlisted the support of the UAW, and although the union had supported the initial implementation of the CAFE standards, it sided with the industry in opposing the proposed changes. As in the timber and coal-mining cases, the union had seen its membership diminish in the decades leading up to the CAFE fight. Between 1979 and 2002 the UAW lost 420,000 auto industry members (Muller 2002). Foreign com-petition and automation had reduced the ranks of the UAW, and union officials were wary of any measure that might further weaken the com-petitive position of U.S. automakers. Although previous job losses had been unrelated to environmental measures, union leaders still viewed environmental action of the sort embodied in the proposed revision of the CAFE standards as a threat to the members' jobs. This economic insecurity is evidenced in UAW's 2002 position paper on the proposed CAFE changes: "The domestic automotive industry has slowed significantly this year, resulting in numerous layoffs. A continuing negative impact on sales and production is expected from the weakening U.S. economy. Given these conditions, it is important that any changes to CAFE not aggravate the challenging economic circumstances facing automakers and their suppliers and result in additional job losses for American workers" (UAW Community Action Program 2002).

The debate over the proposed revisions to the CAFE standards represents another classic jobs-versus-the-environment dispute in which workers, with the support and backing of industry, are pitted directly against environmentalists. In their joint efforts to thwart CAFE changes, union and company officials held a series of rallies in 2002 in communities throughout the Midwest, where auto production facilities are located. Ronnell Coleman, president of UAW Local 597 in Missouri, was quoted by the Associated Press (2002) "We believe in the need for fuel economy improvements, but we don't accept radical plans that could result in a loss of Missouri jobs." Those opposed to CAFE changes ultimately triumphed in the dispute, as the measure proposed by Kerry and Hollings was defeated.

Another recent case offers a slight twist on the jobs-versus-the-environment scenario. Typically unions join the battle to resist any threat of job loss that is allegedly presented by some environmental measure. That these groups should mobilize on this issue is expected, given the concentrated interest that they have in the issue and the level of organization that unionized employees already possess. But a recent dispute between environmentalists and the Teamsters union provides an example of a case in which it is not a threat to existing jobs, but rather the prospect of new jobs, that generates conflict.

In 2002 the administration of President George W. Bush proposed exploratory oil drilling in the ANWR. The oil industry had long sought access to the refuge, where it was estimated that as much as 9.2 billion barrels of oil is located. Sensitive to public opinion on the environment, previous presidents had not strongly advocated for wilderness drilling, but in the younger Bush, the industry found a champion for their cause.

The oil industry has very close ties with the Bush administration, not only because of its strong support for his candidacy, but also because Bush himself is a former industry executive, and many of the positions within his administration have been filled with former industry colleagues.

The proposal for drilling in the ANWR met with the expected resistance from the environmental community, which argued that the relatively small amount of oil to be found in Alaska did not justify the infringement on this wilderness preserve, especially when energy conservation programs could easily create more savings than the extra oil could provide. The Bush administration and allies in Congress, however, empowered by their popularity during the war against terrorism, pushed the need for tapping this domestic oil resource in order to reduce dependence on foreign oil, especially that which came from the volatile Middle East.

Several unions, including SEIU and the Communications Workers of America, joined the environmental community in condemning this opportunistic proposal, however the Teamsters union, along with the building trades, came out in support of the measure (Alvarez and Kahn 2001; Greenhouse 2001). The Teamsters engaged in a particularly ambitious campaign in which union president James Hoffa Jr. lobbied Congress in favor of the proposal in tandem with administration officials. The public justification offered was the number of union jobs that the endeavor would create, which Teamsters officials estimated at 25,000.

In this example, it was not the threat of job loss, but rather the promise of job creation, that motivated union leaders to break with the environmental community. Some, however, offer an alternative explanation for the actions of union officials in this matter ("Teamsters and Turtles in [Temporary?] Tiff" 2002). Since 1989 the Teamsters union has been under federal oversight because of its previous involvement with organized crime. Hoffa's platform when running for the president of the union proposed to end this oversight, which the union is required to finance (Greenhouse 2000). Some have surmised that the Teamsters support for the ANWR proposal stems more from its desire to win favor from the Bush administration for having the oversight removed than for the actual hope of job gains. This is consistent with other Teamsters efforts, such as pressuring the AFL-CIO to endorse more Republican candidates for office, a move obviously favored by the Republican Bush administration. The Teamsters were one of only two unions in the AFL-CIO to oppose increasing political spending by the federation. They did so on the grounds that the AFL-CIO directs too much support toward Democratic candidates. In another move that some view as a cynical ploy to garner the aid of the Bush administration in its effort to have oversight removed, the Teamsters have pledged to contribute one-third of their own political money to Republican candidates, this despite the poor record of most Republican officeholders on issues important to union interests, including some that directly affect the Teamsters.

The Teamsters case again indicates the importance of assessing the role of leadership in any analysis of labor-environmental relations. Some of the inconsistencies found in the pattern of labor-environmental ties can be attributed to the idiosyncrasies associated with particular leaders and the circumstances that led them to power. Prior to Hoffa's ascendancy, the Teamsters were led by Ron Carey, the head of the reform movement Teamsters for a Democratic Union (TDU), and a progressive coalition builder. Carey was removed from office when irregularities were discovered in the financing of his election campaign, allowing Hoffa to capture the leadership position. The TDU has joined with environmentalists in opposing drilling in the ANWR, and had Carey retained his position as president of the union, the Teamsters would likely be in alliance with, rather than opposed to, environmentalists on this issue.

The threat to existing jobs allegedly presented by environmental measures is the most common source of conflict between the labor and environmental sectors. Other economic interests can also play a role in the complex dynamics associated with labor-environmental relations. The prospect of job creation is at least the publicly proclaimed basis for Teamster opposition to environmental-preservation goals in the ANWR. But specific organizational considerations may also come into play. In this case the Teamsters' real agenda may be securing more member or leadership control over the union's operations and ending the costly federal oversight. An obvious question is whether this position is truly in the interest of the union as a whole or if support for ANWR drilling and the political maneuvering involved in offering that support will serve only the interests of the union's leadership. Scholars have long debated the role of leadership in member organizations. Many treat an organization as a single entity with unified objectives, whereas others argue that leaders ultimately have interests distinct from the members they are supposed to represent. This issue will be explored in greater depth in a later chapter.

New Labor Leadership and the Quest for Common Ground: Kyoto and Beyond

As the cases discussed earlier in the chapter demonstrate, the international unions that comprise the bulk of the labor movement in the United States have had varying relations with the environmental movement, and even within the context of that variance, relations with particular environmental organizations have seen their ups and downs. Specific pieces of legislation or policy have affected segments of the labor force in different ways. The effect of environmental policy on particular worker interests, along with other factors, have caused unions to react negatively or positively to environmental measures. This variability in union response has had a complex influence on the AFL-CIO as a federated organization.

The AFL-CIO's primary role is to advance the political interests of workers on the national level. As a federation whose members include most of the U.S. international unions, the AFL-CIO has found itself in difficult situations when members adopt conflicting positions on particular issues. Resolutions passed at national conventions represent the formal position of the federation, but ongoing political developments require that federation leaders develop political strategy on a case-by-case basis. Because of the federation's structure, its leaders tend to refrain from actively advancing goals on which there is not a consensus among the major affiliates, as membership in and the contribution of financial support to the federation on the part of the international unions is voluntary. Most power in the AFL-CIO therefore resides with the international unions, who essentially hold a veto over the actions of the federation. An international union's threat of exit and the potential weakening of the federation forces AFL-CIO leaders to appease any dissenting voice and act only on the lowest common denominator in terms of member union interests.

In most cases, when member interests conflict, the federation will refrain from taking a stance on the issue involved and allow the respective internationals to pursue their own political goals. On occasion, the federation will take a position on a contested issue if it is significant for one union and tangential to another. Environmental issues are often viewed as residing within the tangential category. Thus even when there is member union support for an environmental goal, if a single large member union views it as a threat to its core job interests, the AFL-CIO is unlikely to take a strong position on the issue. Yet federation leaders, who at least symbolically represent the leadership of the labor movement in the United States, can in some ways set the strategic course for the movement and influence the member unions. This capacity is very apparent today, as new AFL-CIO leadership has charted a new course for organized labor in America.

In recent years some dramatic changes have taken place at the national level of the AFL-CIO resulting in a shift in some specific policies and in the overall strategy adopted by the labor movement. In 1995 the first contested election ever for the leadership of the national body was held. The contest was seen by many as a struggle for the soul of the labor movement in America. It pitted John Sweeney, the president of the SEIU, against Thomas Donahue, who was serving as acting president of the AFL-CIO, having moved up from the position of secretary-treasurer when the former leader, Lane Kirkland, resigned in 1995. Sweeney was viewed as the insurgent candidate who called for reforms within the AFL-CIO and a more activist, organizing-focused labor movement. Donahue, although himself a reformer, was more closely tied to the former president and the old guard, which was blamed by some for the dramatic decline in union membership and political influence over the last several decades. A failure on the part of the Kirkland/Donahue administration to respond to global economic change and an aggressive corporate attack on unions was viewed as having contributed to the decline in union density to a forty-year low of just 15 percent of the labor force in 1995. Sweeney won the election and has since brought new energy to the struggling labor movement by advocating the redirection of union resources toward organizing campaigns. Although unions have continued to decline in terms of total percentage of the workforce during Sweeney's presidency, largely because of the ongoing reduction in employment in organized labor's industrial stronghold, the pace of the decline has slowed, and labor has scored some notable organizing victories.

In addition to rededicating itself to organizing, another part of the federation's new revitalization strategy has been to reach out to community organizations and other progressive forces to forge alliances and build united strength. Included in this has been a concerted effort to resolve differences and strengthen ties with the environmental movement. Jane Perkins, a former union officer and president of the national environmental group Friends of the Earth, was appointed by Sweeney to the new executive position of environmental liaison. Perkins was introduced to the labor-environmental nexus as an officer of an SEIU local in Pennsylvania. She was among the people evacuated during the Three Mile Island nuclear accident, and the experience inspired her to seek ways to merge her labor and environmental concerns. Although she met with little success in her efforts to bridge the gap around nuclear energy at the time of the Three Mile Island crisis, her involvement in environmental issues ultimately culminated in her leadership position with Friends of the Earth. After leaving that organization, her union background and environmental credentials made her an ideal liaison between the AFL-CIO and the Washington environmental community, with which she was already familiar. In her position at the AFL-CIO it was Perkins' responsibility first to discern the relationship between diverse member interests and environmental measures and then to develop a common labor position on environmental matters and to seek ways to reach out to the environmental community in order to build alliances.

Recognizing the complex multitude of environmental issues and their inconsistent impact on workers, Perkins selected a single issue on which to forge a coherent labor position and to attempt to devise a common strategy with environmentalists. The issue, on which unions and environmentalists were extremely divided, was global climate change. By the 1990s there was a growing scientific consensus that human activity, especially the extensive burning of fossil fuels since the start of the industrial revolution, had resulted in a buildup of carbon dioxide and other gasses in the atmosphere. These gases trapped the sun's heat, creating a greenhouse effect on earth and disrupting the climate generally. Average global temperatures were found to be increasing over the last century, with several years during the 1990s being the hottest on record. Recognizing the seriousness of this problem, virtually all of the major national environmental organizations had made climate change a priority issue.

The fact that the use of fossil fuels was the central concern meant that the issue of climate change could have dramatic ramifications for the economy and jobs. By the time Perkins was formally appointed to her position at the AFL-CIO, unions had already taken a firm position against an international treaty being forged to address global climate change on the grounds that it threatened jobs in the United States. The main basis for labor's opposition to the treaty stemmed from the particular approach to the problem adopted by the world community. Because it is a global problem, climate change requires a global solution. A framework for addressing the problem on an international level was hammered out in Berlin in 1996. In developing that framework it was determined that initial reductions in greenhouse gas emissions should take place among the industrialized nations. This approach was justified first on the grounds that the industrialized nations were responsible for the vast majority of greenhouse gas emissions and second on the grounds that the wealthier industrialized nations could better afford to adopt pollution reduction measures. The less developed nations would not be expected, under the proposed approach, to take action until some unspecified point in the future.

This two-tiered approach served as the sticking point for organized labor. Union leaders feared that the more stringent requirements faced by the United States with regard to greenhouse gas emissions would facilitate job flight to the less developed nations with their more relaxed requirements. Soon after the establishment of the treaty framework in Berlin, union opponents of the treaty mobilized federation opposition. According to an AFL-CIO Executive Council statement (1997): "The exclusion of new commitments by developing nations under the Berlin Mandate will create a powerful incentive for transnational corporations to export jobs, capital, and pollution, and will do little or nothing to stabilize atmospheric concentrations of carbon. Such an uneven playing field will cause the loss of high-paying U.S. jobs in the mining, manufacturing, transport and other sectors."

The federation formally opposed the final treaty because it included the differential treatment of developing nations. The Kyoto Protocol, named for the city in which the final treaty was forged in 1997, called upon industrialized nations to reduce their combined greenhouse gas emissions to 5.2 percent below 1990 levels by 2012 while placing no corresponding requirements on less developed nations. The treaty was signed by the United States under President Clinton in December 1997, but despite the president's signature, the United States would not be bound by the treaty until it was passed with two-thirds support from the Senate. Environmentalists called for the treaty's passage, but opposition from labor and industry ensured that the treaty would never receive the Senate's approval.

It was these sharp divisions between unions and environmentalists on such a serious concern as global climate change that led Perkins to make this issue the focus of her efforts. To address this most pressing issue, Perkins organized a series of meetings between labor and environmental leaders to discuss their differences and to seek a common position on global climate change. Although some international unions had ongoing relations with environmental organizations and ad hoc projects have been carried out by the AFL-CIO and some national environmental groups in the past, the meetings organized by Perkins were the most comprehensive effort ever undertaken to unite these movements at the national level. Not all AFL-CIO affiliates were enthusiastic about the prospect of working with the environmental community on this issue. In particular, the United Mine Workers, still reeling from losses that resulted from the Clean Air Act, were vocally opposed to any concessions on the matter of climate change.

The Climate Change Working Group, as the effort organized by Perkins came to be called, held a series of meetings beginning in 1998 that included high-ranking representatives of almost all the major international unions and the national environmental organizations. These meetings went on for several years, during which the participants painstakingly hammered out a set of principles regarding how climate change could be addressed while protecting the interests of U.S. workers.

Although some progress was made in terms of laying out basic principles for a common position, achieving agreement on the intricate details necessary for any such plan proved elusive. Continued resistance from several unions within the federation ultimately forced the end of the AFL-CIO's direct sponsorship of the effort. Several of the more committed unions and environmental groups went on, however, to form the Blue-Green Alliance to continue their work. Perkins, now based in the George Meany Center, the AFL-CIO-sponsored labor education institute, continues to facilitate dialogue between unions and environmental organizations on this issue.

Conclusion

The cases of labor-environmental conflict and cooperation over the last several decades discussed in this chapter provide some sense of the trajectory of labor-environmental relations during that time period. It is apparent that several different political and economic factors may have an impact on the types of relations that exist between labor unions and environmental advocates. There is also clear variation among different unions in the way in which they approach environmental issues. This variation can be explained in part by the different ways in which distinct segments of the workforce are affected by environmental policy. For example, to some extent it is clear why timber industry workers would feel threatened by forest preservation policies. Health concerns also make it clear why chemical workers who handle toxic substances find common cause with environmentalists. Yet a simple analysis of interests and environmentally related employment effects cannot tell us when cooperation or conflict between labor and environmental organizations will emerge. The quality of the relations between these organizations appears to hinge on a host of variables.

Because of the complexity of the interaction of the labor and environmental sectors around different issues, it is impossible to offer a simple characterization of the labor-environmental relationship. Different unions and environmental advocates have adopted divergent approaches to the range of issues that have emerged over the last several decades. At any given moment on any given issue there are some unions and environmentalists that are working together and others who are at odds. Yet it is still possible to detect shifts in the overall mood at the national level.

In general, relations appeared good at the start of the contemporary environmental movement. Even before that time, unions tended to be supportive of environmental concerns. But by the mid-1970s, union skepticism of the motives of environmental organizations was growing. The energy crisis and recession of the 1970s encouraged many unionists to focus on their traditional concern with employment and living standards, and the calls by some environmentalists for a zero-growth economy clashed with this basic orientation. The emergence of groups designed to build labor-environmental links, such as EFFE, and the high-level contacts that were made between union and environmental leaders during this period can be interpreted as an effort to maintain ties in the face of these deteriorating relations.

This effort to counteract the growing tensions around such issues as nuclear energy and to salvage positive ties between unions and environmentalists met with some success in the early 1980s in large part because of the mutual opposition of these two groups to the Reagan administration and its conservative antilabor, antienvironmental orientation. In addition, a growing focus on toxic substances and workplace and community environmental hazards provided more common ground for these sectors. But conservatism on the part of most national labor and environmental leaders at the time led both sides to take defensive positions around their relatively narrow concerns, leaving little room for cooperative efforts around a broader progressive agenda.

High-profile conflicts emerged again later in the 1980s when timber industry workers and miners both faced significant job threats generated by environmental measures. These conflicts once again reinforced the notion of a jobs-versus-the-environment trade-off, and relations between unions and environmental organizations soured. But by the early 1990s there was renewed hope for a broad sustained labor-environmental alliance. Environmental-justice groups and the established environmental organizations that responded to their critique regarding the mainstream movements narrow wilderness focus brought the social impact of environmental hazards into view, allowing for strengthened ties with some segments of labor throughout the 1990s.

Free trade also emerged in the 1990s as an issue around which unions and many in the environmental community could come together. Despite the fact that several environmentalists defected from the anti-NAFTA alliance and came out in support of the trade agreement, opposition to the WTO has since served as an important rallying point for labor-environmental cooperation. And there are signs that at least at the top levels, union and environmental leaders are making concerted efforts to build a sustainable alliance. The current leadership in the labor movement is much more inclined toward alliance building than their predecessors, and there has been considerable outreach by labor organizations to the environmental community in recent years. Some issues continue to generate conflict between environmentalists and segments of the labor community, but on the whole relations between the two at the beginning of the twenty-first century can be seen as on the upswing.

4
Labor and Environmental Relations: State Cases

Based on the interest assessment offered in chapter 2 and the history of labor-environmental relations at the national level reviewed in chapter 3, it is clear that many factors influence the quality of the relations between unions and environmental advocates. A superficial assessment suggests that the relationship between these two sectors is characterized by conflict and that this conflict stems from a basic clash of interests. Environmental measures harm employment, according to this perspective and environmentalists working to protect the environment will inevitably clash with unions who are fighting to protect jobs. But further scrutiny reveals that there are some fundamental flaws in this conventional wisdom. First, economic analyses show that the effect of environmental protection on employment is far more complicated than initially believed. On the whole macroeconomic analysis reveals that the conventional view is simply not correct: Environmental protection actually creates jobs. When we consider the historical evidence, we see that there are instances in which unions and environmentalists have worked together to achieve the mutual goals of environmental protection and job creation.

But the real question is "whose jobs?" Who may lose their jobs as a result of environmental measures, and what jobs will be created? When we ask that question, we can see the ground on which the conventional argument stands. Some workers may lose their jobs as a result of environmental measures even while others gain a new source of employment. We can expect that those workers who lose jobs, such as those employed in the Northwest's timber industry or those working at marginally profitable tanneries in Fulton County, New York, will engage in struggles against environmentalists. Yet job losses from the enactment of environmental protection policies are not nearly as great as one might expect based on the attention that these struggles receive. Environmentally related job loss is a mere blip in the overwhelming cacophony of economic fluctuation in which tens of thousands of workers may lose their jobs in a single corporate restructuring. Much more is needed to understand when labor-environmental cooperation or conflict will occur than a simple examination of how employment is affected by a particular environmental policy.

At the very least, we must also consider other ways in which the interests of workers are influenced by environmental measures, such as those that protect them at the workplace. The positive relations between chemical industry workers and environmentalists can be understood at least in part in terms of such interests. Threats to jobs resulting from environmental regulations are counterbalanced in many instances by concerns about health and safety. In other cases political interests influence intermovement relationships. In some instances mutual support for prolabor, proenvironment political candidates can unite disparate forces in the labor and environmental movements. But even this expanded analysis of interests raises many questions. When multiple interests are involved, how do unions come to prioritize one issue over another? How do environmentalists select issues that may or may not mesh with the interests of workers? What ideological predispositions shape their approach to these issues? What role do leaders of environmental and labor organizations play in establishing cooperative ties between the two?

The bottom line is that interests do matter in terms of how two organizations will get along. But how those interests are defined is the product of many forces. The historical evidence regarding relations between unions and environmentalists at the national level reveals some of the other factors that come into play. But focusing exclusively on national-level issues and high-profile cases masks what may be happening at the state and local levels. There is some evidence that community-based alliances are increasingly common (Brecher and Costello 1990). In recent years the national AFL-CIO has been encouraging this kind of outreach on all levels. A full understanding of labor-environmental ties requires that we look beyond the cases that make national headlines.

But local community ties between environmentalists and unions are difficult to gauge. The presence of actors from these two movements at the community level varies dramatically from community to community, and unless a local issue relevant to both develops, interaction between these two sectors is likely to be sporadic. In contrast, unions and environmental organizations are active at the state level across the nation. This is also the level at which many relevant policy decisions are made. The continuous development of policy at this level allows for ongoing interaction between unions and environmentalists, thus allowing ongoing relationships between the two to develop. Interactions at the state level can therefore provide a fruitful basis upon which to examine intermovement relations in more detail.

There are some systematic data regarding labor-environmental ties on the state level. These data derive from a survey of state labor leaders conducted in 1997–1998 (Obach 2002). In that survey, the leaders of state labor federations shared their views on the quality of their relations with environmental advocates in their state. Of those surveyed, 64 percent reported that their relations with environmentalists were either "good" or "very good," 26 percent said relations were "fair," and only 10 percent identified their relations as "poor" or "very bad." These ratings of labor-environmental relations at the state level contrast with the impression left by most media reports and scholarly studies, which tend to focus on instances of conflict.

In addition to the generally positive characterization of labor-environmental relations at the state level in the 1997–1998 survey, almost all union representatives who responded placed at least some importance on environmental issues, with almost half identifying them as "very important." The importance given to environmental issues was further evidenced by the fact that 87 percent reported that their federation had taken a position on an environmental issue. As could be anticipated, the most common areas of agreement with environmentalists involved issues related to the workplace, such as the right of workers to know the dangers associated with substances used in the workplace, the in-plant environment, and restrictions on the use of toxic substances. The more general protection of air and water quality were also commonly cited areas of cooperation, as were international-trade issues, which have both environmental and job implications.

But although the state level union respondents were able to offer many examples of agreement and cooperation with environmentalists, not all sentiments in regard to environmental protection expressed in the survey were positive. Eighty-two percent of those surveyed reported that they believed that environmental protection can pose a threat to the interests of workers in at least some instances. The most common areas of disagreement with environmentalists cited were issues related to forest protection, air quality protection, restrictions on development, and bottle bills. In each instance there was a perceived threat of job loss. The loss of coal-mining jobs was the most commonly cited example, but job loss was also identified in the timber industry due to forest protection, in construction due to restrictions on development, and in manufacturing.

These survey data are useful, but there is still a need for a closer examination of the intricacies of the interactions between labor and environmental actors and the complex forces behind the dynamics that surround such interactions. For this we need a qualitative analysis of some actual organizational interactions. We need to speak with the people who are involved in these processes, both those engaged in conflict and those who are working together. The rest of the analysis presented in this book is based on such cases. In the chapters that follow I describe the experiences of dozens of labor and environmental activists and leaders from five states, Maine, New Jersey, New York, Washington, and Wisconsin. I share their perspective on these issues, while also offering my own analysis of the forces that shape intermovement relations. Before delving into these issues, I will provide an overview of the political and economic conditions in each state at the time of the study and briefly review the status of labor-environmental relations.

Maine

Maine is a primarily rural state with almost half of its population concentrated in the southwestern region. As in much of the rest of the nation and all of the states considered here, service is the primary employment sector. Manufacturing still employs about 15 percent of the workforce, however, and it is still the largest contributor to the state's gross product. The pulp and paper industry of the northern portion of the state is the dominant manufacturing industry, and much of the northern forest land is held by large paper companies.

Politically, the state is very independent, with registered independents outnumbering members of both the Democratic and Republican Parties. The Portland area and the southern portion of the state generally have a reputation for being very liberal. People in the more rural northern parts of Maine are more conservative and tend to be resentful of the influence that comes with the larger population concentrated in the south. There is a general suspicion among residents of northern Maine of people "from away," and to some, southern Maine is an outpost for migrants from out of state. Large paper and timber interests in the northern region have a great deal of political influence because of the number of people they employ, their economic significance in the state, and the organized lobby they have created.

The progressive community in the southern portion of the state is well organized politically. Liberal community organizations hold regular meetings (including representatives of organized labor and environmental organizations) to coordinate electoral strategy and to engage in issue campaigns. They have successfully recruited and run several candidates for state office in addition to initiating legislation.

At the time that the research for this book was conducted, both houses of Maine's legislature were controlled by the Democratic Party, in part because of the success of the progressive coalition in the state's southern region. Although the governor at the time was an independent, many liberals complained that his policies favored big business. Leaders of the state labor federation expressed decidedly liberal leanings. The state's paper industry, however, is heavily unionized, and the generally more conservative United Paperworkers International Union (which merged with the OCAW in 1999 to become the Paper, Allied Industrial, Chemical, and Energy Workers International Union, or PACE) exercises a great deal of influence within the state labor federation. Some PACE locals are very progressive politically, in part resulting from a highly contentious strike in Jay in the late 1980s, which radicalized the local membership.

During that time union supporters gained control of the local government and used their position to impose strict environmental requirements on the local paper mill. However, although the paperworkers' union may be undergoing some changes, on the whole, the PACE leadership is wary of environmentalists and their efforts to impose environmental restrictions on the paper industry and to regulate forest practices in ways the union believes would threaten the supply of wood for making paper.

This skepticism regarding environmentalism on the part of the paperworkers' union is reflected in and aggravated by the presence of an antienvironmental quasi-worker organization, the Pulp and Paperworkers Resource Council (PPRC). This industry-funded organization was founded on the West Coast in 1992 during the disputes in that region over the protection of the endangered spotted owl. According to the organization's own history, paper workers, frustrated with the threat to their jobs posed by overzealous environmentalists and lacking the additional resources needed to challenge them through their unions, approached management seeking funds to create a workers' organization designed to address "fiber supply, forest practices, The Endangered Species Act, and our environment . . . so that we may influence legislation that affects our jobs" (PPRC n.d.).

Although it is alleged to be a grassroots worker organization, some claim that the PPRC is part of the "wise use" movement, an industry-generated effort to create the appearance of popular opposition to environmental regulation (Echeverria and Eby 1995). Such industry-generated organizations have been known to compensate workers with time off or additional pay for political activities in support of industry-backed legislation. "Grassroots" PPRC leaders are paid by their employing companies and given additional pay and release time for their roles in the organization.

The PPRC expanded to the East Coast in 1994 creating a Northeast regional chapter that enlisted several union locals from the paper industry in Maine. Although several union leaders at the state level consider the PPRC to be an industry front group, a modicum of support for it at the grassroots and local levels has in some ways inhibited the ability of state leaders to take more proenvironmental positions.

The environmental community in Maine is divided between radical and more moderate factions. The Natural Resources Council of Maine (NRCM) represents the more mainstream environmental movement, along with the Maine Audubon Society. These are staff-driven advocacy organizations with a sizable yet largely inactive membership. The Maine Greens and the Forest Ecology Network represent the more radical and more grassroots-based branch of the state's environmental community, along with a loose collection of Earth First! activists. The Sierra Club also has a political presence in Maine, and although it is also considered to be part of the mainstream environmental movement, the state Sierra Club chapter has sided with the more radical element on some key issues.

There have been some classic labor-environmental conflicts in Maine through the years. For example, unions and environmentalists clashed over a proposed hydroelectric dam in the 1980s. But the 1990s saw some important cooperation on toxic emissions. Unions supported environmental legislation to reduce the amount of toxic discharge into the state's rivers. There has also been some cooperation between unions and the mainstream environmentalists on forest practices in the state. But close ties between labor and the mainstream environmental movement have highlighted the divisions between the moderate and radical environmentalists in the state. The main fissure between these elements centers on forest practices and in particular a voter initiative in 1996 that would have imposed a ban on the practice of forest clear-cutting. The Greens, attempting to increase their profile in the state, began running candidates for statewide office during the 1990s. Jonathan Carter, a Greens activist, ran for Congress in 1992 and then for governor in 1994, receiving 6.5 percent of the vote in the latter contest. Attempting to build on that momentum, the Greens began a petition drive to have the clear-cutting ban placed on the next ballot.

Forest policy had been a serious point of contention in the state for years. Woodlands cover over half of the state, yet only 5 percent of the state's land is publicly held, one of the smallest ratios of publicly owned land in the nation. There had been calls through the years for more public holdings and even the establishment of a national park to prevent the liquidation of forests under private industry control. Clear-cutting had increased in the 1990s, and studies had demonstrated that the forestry practices in use were not sustainable in the long term. Legislative efforts to reform logging practices in the state in the early 1990s had failed, leaving many in the environmental community frustrated. The proposed clear-cutting ban was the Greens' answer to this perceived threat to the state's forest environment.

Under Maine law, a measure that achieves a given number of signatures first goes before the legislature for a vote. If it fails, it is then placed on the ballot for a binding public vote. The Greens' petition drive was successful in obtaining enough signatures to require legislative consideration, but the resulting measure did not pass in the legislature. However, there was a good deal of popular support, and there was a real chance of passage in the November election. The paper industry, as might be expected, opposed the measure and began applying pressure to undermine the ballot initiative. Political opponents took advantage of a rarely used provision in the law whereby an alternative measure can be placed on the ballot along with the original initiative. They sought to undercut the clear-cutting ban by offering a compromise alternative that came to be known as "the Compact." Political leaders, paper industry officials, labor leaders, and members of the mainstream environmental community forged the alternative, which proposed to reform logging practices while stopping short of a clear-cutting ban. Thus, voters were faced with a choice between the clear-cutting ban supported by the more radical environmentalists, the Compact, which had the backing of industry, labor, and the mainstream environmental groups, or none of the above. To further add to the divisions, a small but significant libertarian faction in the state mobilized to oppose both policy alternatives on the grounds that the state should not regulate the use of private property.

Greens leaders attacked the Compact as a sellout to industry. The fact that the mainstream environmental organizations, including the NRCM and Maine Audubon, supported the Compact lies at the root of the divisions within the state's environmental community. The more radical environmentalists charged that the environmental supporters of the Compact had fallen under the control of individuals affiliated with the paper industry who sat on their boards of directors or contributed money to their organizations.

The Compact received the most votes in the 1996 referendum, but it did not receive a majority. According to Maine law, this meant that the Compact alone had to be offered again in another round of voting the following year. To the frustration of labor and the mainstream environmentalists, in this second round of voting the Greens and the more radical wing of the environmental movement joined forces with the libertarians to oppose the Compact altogether on the grounds that it would institutionalize ecologically unsound forestry practices. The Compact failed in the next round, and divisions among environmentalists ran deep.

Although there is some animosity among the environmentalists, relations between organized labor in the state and the mainstream environmental community are currently quite good. Given the small state population, contact among all political actors is fairly routine. The progressive coalitions, which incorporate members of the various mainstream organizations, provide an additional forum for homogenizing interests and forging a common agenda. The more radical environmental organizations, however, tend not to be represented in these coalition organizations. Some members of the mainstream groups view these organizations and those who belong to them as mavericks who are too narrowly focused on their environmental agenda. Although relations between the mainstream and more radical factions of the environmental community have improved somewhat since the divisive clear-cutting issue, the Greens are still not well integrated into the otherwise tight progressive community.

Wisconsin

Wisconsin has a very significant manufacturing base, which includes metal and industrial-machine production. These industries are located primarily in the Milwaukee area and the southeastern part of the state, although there is also a significant paper-manufacturing industry in the northern region. Many of these industries are highly unionized, giving Wisconsin an above-average rate of unionization, which stands at approximately 19 percent. Although many regions in the United States have lost a great deal of manufacturing employment (and Wisconsin has lost some employment in certain industries), Wisconsin actually experienced some gains in manufacturing employment in the 1990s, and the overall rate of unemployment in the state has been exceptionally low.

There is a great deal of environmental activity in Wisconsin. National organizations such as the Sierra Club and the Audubon Society have very large statewide memberships and several active local groups. In addition there are active state-based environmental organizations, the two most significant of which are Wisconsin's Environmental Decade and Citizens for a Better Environment.

Wisconsin's labor movement has had a close relationship with the state's environmental community for almost twenty years (Obach 1999). This relationship developed in 1981 when the national AFL-CIO's Industrial Union Department and the Sierra Club, along with other national environmental organizations, launched an effort to build labor-environmental networks at the state level around the country. This move was inspired by the election of Ronald Reagan and the dual threat posed to the EPA and OSHA by the Republican antiregulatory agenda. Labor and environmental leaders from around the nation were summoned to a meeting in Chicago to establish these OSHA-environmental networks and to coordinate a strategy in defense of these valued agencies.

Dominick D'Ambrosio, then president of the Allied Industrial Workers International Union (AIW), which had its headquarters in Milwaukee, attended the Chicago meeting and returned to Wisconsin to set up the Wisconsin OSHA-Environmental Network (later the Wisconsin Labor-Environmental Network, or WLEN). He was successful at recruiting a number of prominent labor and environmental leaders in the state, including Tony Earl, a state senator and respected member of the environmental community who would cochair the network along with D'Ambrosio (the precedent was then set to have cochairs head the network, with one representative of labor and one from the environmental side). Prior to the establishment of the network, there had been little contact between unions and environmentalists in the state.

Although participation in the network was widespread, including almost all of the major statewide environmental organizations and many of the larger unions in the southern half of the state, membership among labor and environmental organizations was not universal. Paperworkers based in the north signed on only later in WLEN's existence, and the building trades were not represented directly at all. Nonetheless, statewide labor representatives and those from central labor councils that included building-trades unions incorporated their concerns into the network's agenda.

WLEN functioned very well and scored a number of legislative victories, the most notable of which was the passage of right-to-know legislation in the state. As discussed in chapter 3, right-to-know laws give workers and members of the community access to information regarding the toxic substances being used in local manufacturing facilities, enabling workers to better protect themselves from hazards on the job while also giving environmentalists information useful in their efforts to protect the natural environment.

Although WLEN members were able to find many issues of common concern on which to work, not every issue generated the immediate support of every affiliate of the network. In some cases compromises needed to be hammered out, while in others the search for agreement was fruitless. For example, in the late 1980s, environmentalists in the state sought to institute a deposit program for cans and bottles. This "ban the can" program was very much opposed by unionized workers in the can-making and brewing industries on the grounds that it presented a threat to jobs, a claim that environmentalists denied. This classic jobs-versus-the-environment dispute was a point of contention within the organization for several years. Finally a compromise was reached whereby both sides agreed to support an aggressive recycling program, but no deposit legislation. With the active support of labor, the recycling legislation was implemented.

On other issues, however, no compromise was possible. In the early 1990s environmentalists sought to impose a "toxic freeze," limiting the discharge of toxic substances to their levels at that time. Labor opposed this effort, again on the grounds that it posed a threat to jobs. On this issue, no compromise was possible, and environmentalists eventually dropped their plans in the face of overwhelming labor opposition.

By the late 1990s, after almost twenty years of active labor-environmental cooperation, WLEN entered a state of dormancy. Key personnel changes destabilized the organization. The AIW had served as the base of operations for WLEN, and Ken Germanson, a staff member at the AIW, had carried out much of the routine work involved in maintaining the network. In 1993 the AIW merged with the United Paperworkers International Union and closed its Milwaukee office. Germanson retired and withdrew from WLEN, and although labor leadership was formally passed on to an activist with the steelworkers union, meetings became less regular and eventually ceased altogether.

Despite the decline of WLEN, labor-environmental relations are still cordial in Wisconsin. The current president of the state AFL-CIO was a long-time participant in WLEN, and the two sides describe relations between them in positive terms. There have been sporadic coordinated activities around some issues, such as free trade, and some statewide events sponsored by progressive organizations bring members of labor and environmental organizations into contact with one another.

One issue has threatened to create divisions in recent years: a proposed copper mine near the headwaters of Wisconsin's Wolf River. Environmentalists and Native American organizations oppose the mine because of the threat it presents to the health of the river. In one incident, which heightened labor-environmental tensions, the president of a Steelworkers local in Milwaukee appeared in a television commercial sponsored by the mining company endorsing the proposal for its job creation potential. Activists within the labor community quickly mobilized to counter the appearance of widespread union support for the mine and denounced the individual in the commercial. Although the Steelworkers district council did go on record in support of the mine proposal, numerous other unions from around the state voiced public opposition to the mine not only on environmental grounds, but also because of the poor labor record of the parent company of the firm engaged in the project.

New York

New York is the second most populous state in the nation. Although over 40 percent of the state's workforce is employed in services, New York still has a sizable manufacturing base. Despite the fact that manufacturing employment in the state has declined significantly in the last twenty years, manufacturing still employs 11 percent of the state's workforce. Extractive industries represent a small percentage of the overall state workforce. New York was the most highly unionized state in the nation, with a union density of almost 27 percent in the 1990s, giving organized labor a strong voice in state politics.

Politically, New York has a long liberal tradition. From Franklin Roosevelt to Mario Cuomo, it has had several governors who have served as the leading voice of liberal politics in the nation. However, the general rightward political shift in the United States that began in the 1980s was also reflected in New York politics. At the time of publication, the state is headed by a Republican governor and the Republican Party holds a majority in the state senate. Democrats continue to control the state assembly.

Several environmental organizations are active in the state, including state chapters of national organizations such as the Audubon Society, the Sierra Club, and the League of Conservation Voters. These are politically active organizations in terms of advocacy, most of which is carried out by a small number of professional or volunteer staff, although some have functioning local groups. Other state-based environmental organizations are also very active politically including the Citizens' Environmental Coalition (CEC) and Environmental Advocates (EA). EA operates like a typical staff-driven professional organization, which primarily monitors and advocates on environmentally related legislation. CEC focuses more on industrial pollution and issues of environmental justice.

Relations between labor unions and environmentalists in the state are positive. Ties between environmentalists and union workers were first established at the local level. Workplace and community environmental health risks in two localities in the state provided the initial impetus for labor-environmental cooperation. The first was the Love Canal incident in Niagara Falls in 1978, in which an entire neighborhood was found to be contaminated with toxic waste. Among the residents of the community were members of the UAW, and the union offered support to the community and environmental organizations that organized around that incident. An alliance of unions and environmental organizations grew out of this development in western New York.

In another case in central New York, workers and environmentalists were brought together in opposition to a company that had exposed workers to dangerous levels of uranium, while at the same time creating an environmental hazard through the dumping of the radioactive waste. The CEC's Anne Rabe had close ties with the local Committee for Occupational Safety and Health (COSH), a labor organization focused on health and safety issues. While COSH was investigating the exposure of the company's workers to radioactive substances, a uranium dump created by the same company was discovered. The CEC worked with COSH and the exposed workers, and out of that cooperation grew a second local labor-environmental network.

The local ties that originally developed based on these incidents were expanded and formalized in 1989 when a statewide conference was held to bring together unions and environmentalists on a larger scale. Out of this the New York State Labor and Environment Network was formed. The network is made up of labor, environmental, and community organizations from throughout the state. The CEC is the most active environmental organization involved in the network. Labor representation is quite broad, including the United Steelworkers, the UAW, various public employee unions, the state AFL-CIO, and a host of COSH organizations. The network holds annual conferences to develop common strategies on issues of mutual concern, and its board of directors meets regularly throughout the year to share information and coordinate action.

As is common in every state, isolated conflicts between labor and environmental organizations routinely crop up around specific issues. Recently tension between environmentalists and unionized workers has reemerged over the Indian Point nuclear power plant. Environmentalists renewed their calls for closing down the plant in the wake of the September 11, 2001, terrorist attacks on New York City and Washington. Citing the vulnerability of the facility and its proximity to major population centers, environmentalists contend that it is a disaster waiting to happen. As might be expected, the unionized employees at the plant reject this charge and cite the severe economic repercussions, including job loss and energy shortfalls, should the plant be decommissioned. The regional labor federation, which maintains cooperative ties with some environmental organizations, has sought to diffuse the conflict between labor and environmentalists over the Indian Point facility, but agreement or any compromise remains unlikely.

The building trades, a significant labor presence not represented in the labor and environment network, have also had some conflicts with environmental advocates over various development projects. Despite these conflicts and the building trades' lack of involvement with the network, there are also examples of unilateral cooperation between the building trades and environmentalists in the state. The Audubon Society in particular has engaged in cooperative efforts with the building trades on projects designed to protect the waters of Long Island Sound. In general, as a result of the work of the state labor and environment network, relations between labor and environmental organizations in New York are positive, but isolated areas of tension remain.

New Jersey

New Jersey is the most densely populated state in the nation. As in every state, the service sector in New Jersey employs the highest number of people, yet New Jersey still has a significant manufacturing sector. The northern half of the state is highly industrialized, and although the state lost a number of manufacturing jobs throughout the 1980s, over 13 percent of its working population is still employed in manufacturing. New Jersey leads the nation in chemical manufacturing; industrial machinery and food processing are also significant manufacturing industries in the state. There is very little in the state in the way of extractive industries such as timber.

The state has a relatively high rate of unionization, with over 21 percent of the workforce belonging to a union. The labor movement in the state, however, maintains a division that has existed since the merger of the AFL and the CIO in 1955. The industrial unions of the CIO in New Jersey never formally joined in on the national merger of the two primary labor federations. Although some industrial unions are affiliated with the state AFL-CIO, the formal state labor federation associated with the national federation, many that were historically affiliated with the CIO work primarily with the New Jersey Industrial Union Council.

Whereas the New Jersey AFL-CIO tends to focus on narrow, traditional union issues such as jobs and wages, the New Jersey Industrial Union Council is more active in a broader range of issues affecting workers, including the environment. Leaders of the Industrial Union Council are also active supporters of the New Jersey Work Environment Council, a labor organization that actively strives to build alliances with environmental organizations to avoid jobs-versus-the-environment conflicts and to improve health and safety conditions at the workplace.

There are several active statewide environmental organizations operating in New Jersey. Both the Audubon Society and the Sierra Club have large state chapters that employ a professional staff, including lobbyists. A state branch of the national organization Clean Water Action, locally called the New Jersey Environmental Federation, also has a professional staff and conducts a canvass operation. It has organizational affiliates (including some labor unions) as well as individual members, and it maintains a separate political-action committee that does explicitly political work. The Environmental Federation works on a wide range of environmental issues and is the most active environmental group in the state in terms of working with labor organizations.

Clean Ocean Action is a separate state-based organization that spun itself off from the American Littoral Society (which also has a state organization). It is primarily based on organizational affiliations, but individuals can also sign up as members. The group focuses specifically on ocean-related issues. Clean Ocean Action was at the center of one of the state's most divisive labor-environmental conflicts, which involved the ocean dumping of dredged toxic material from the state's biggest harbor. Although the issue was eventually resolved, the conflict reflects the poor relations that exist in New Jersey between environmentalists and at least some segments of the labor community.

Of all the states examined in this study, labor-environmental relations are worst in the state of New Jersey. Yet although relations between the state AFL-CIO and environmental organizations are poor, individual ties between some unions and environmental organizations are quite positive. The New Jersey AFL-CIO, because of its narrow focus on job-related issues, has come into conflict with environmentalists on several occasions and generally shuns coalition efforts. A number of unions in the state, however, in particular the OCAW (which is now PACE, as noted earlier in the chapter), have been very active on environmental issues. OCAW leaders have worked closely with environmental organizations in a number of instances, and they, and several other union locals are formally affiliated with some of the environmental coalitions in the state. The New Jersey Work Environment Council also serves as an important bridge between unions and the environmental community. Although no statewide labor-environmental network exists in New Jersey, these ties have allowed for some cooperative labor-environmental efforts, despite tensions between environmental organizations and the state AFL-CIO.

Washington

The state of Washington has large aerospace, aircraft, and shipbuilding industries. Pulp and paper are also manufactured in the state, which has a significant lumber industry as well. Washington is in fact a major producer of lumber, second only to Oregon. This is due to the fact that half the state is forest covered, much of which is used for commercial purposes. Limits on forest cutting stemming from the spotted-owl disputes, in addition to the overcutting that had occurred in the decade previous to those disputes, disrupted the timber industry and its workers. This had a significant impact on several small timber-dependent communities.

Much of the state's population is concentrated in the Puget Sound area. The manufacturing industries are located in this part of the state, in addition to a significant fishing industry. Construction employment is higher in Washington than in most states in part because of the population growth that the state has experienced in the 1990s. Over 6 percent of the population is employed in construction. Washington's economy is also highly dependent on international trade. Almost 20 percent of the state's workforce is unionized.

There are many environmental organizations active in Washington. The Washington Environmental Council (WEC) is a professional environmental group with eighty-five organizational affiliates composed primarily of environmental and community organizations. Its professional staff focuses on monitoring and advocating on environmentally related legislative issues. Another group, People for the Puget Sound, focuses on protecting the waters of the Puget Sound from dangers such as industrial discharge, oil spills, and sewage leaks. Its strategy involves forming alliances with local businesses such as the shellfish industry, which relies upon the resources drawn from the sound. There are also active chapters of both the Sierra Club and the Audubon Society that have professional staff, including lobbyists.

Relations between unions and environmentalists in Washington are fair at present, although they have undergone significant ups and downs over the last thirty years. Labor's approach to environmental organizations paralleled relations between the two at the national level during the 1970s. The labor movement in Washington was largely on board with environmental causes in the early 1970s when the new wave of environmental mobilization was getting underway. The Washington State Labor Council was among the founding members of the WEC. Tensions between labor and environmental organizations in the state rose in the 1970s, however, as building-trades workers began to view environmental opposition to nuclear power, highway expansion, the Alaskan Oil Pipeline, and other major construction projects as a threat to jobs. Labor organizations pulled out of the WEC in the mid-1970s, and relations between labor and environmental organizations in the state remained poor for many years thereafter. Some reconciliation took place in the early 1990s as labor and environmental organizations worked together on issues involving growth management. Environmentalists even took the reciprocal step of supporting the building trades in efforts to defend prevailing wage requirements (Rose 2000, 146–159).

Yet despite some headway, tensions between labor and environmental organizations in Washington flared again over timber-related issues in the 1990s. These tensions have not fully dissipated. One might also expect that the pressure posed by the state's rapid population growth would facilitate conflict around development issues between the environmental community and the sizable building-trades unions. In fact these issues may explain the poorer-than-average labor-environmental relations in the state. These relations are likely to be tested further, since some species of salmon native to western Washington have been designated as threatened, necessitating that stringent measures be taken to protect their habitat. This could have significant repercussions for the building trades in particular.

Despite these areas of tension, some of the leaders in the state labor movement, who rose up out of the building trades, are very sympathetic to environmental causes. Even when labor's ties to the environmental community have not been close, unions themselves have initiated action against employers who have violated environmental standards. The forward-thinking union leadership in the state has generally adopted a broader interpretation of labor's role than union leaders in many other states. The more encompassing strategy pursued by these leaders has brought them together with environmentalists on some occasions, including collaborative work designed to create employment opportunities for displaced timber industry workers in habitat restoration.

International-trade agreements have also served as a force to unite the state labor and environmental communities. Washington's unions worked closely with environmentalists against NAFTA in the early 1990s. As noted in chapter 3, Seattle was also the site of the massive labor and environmental demonstration against the WTO in 1999. Although this was a national demonstration that brought activists from all over the country and even from some other nations, state labor and environmental activists worked together to coordinate protest activity. Even though there are no formal state-based labor-environmental networks in Washington at present, there continues to be some irregular informal contact between union and environmental leaders on specific issues.

Conclusion

In each of the states examined in this chapter, the interactions between unions and environmentalists provide some insight into how relationships develop between them and the factors that influence that development. In the following chapters, the central themes that have emerged in this examination will be considered in detail. As will be seen, the core interests of the organizations and actors provide only part of the story. What becomes most clear is that intermovement relationship development is a dynamic process whose direction is contingent on many factors. Although the basic interests of workers in different sectors and the core goals of environmentalists can provide some ideas about the way in which intermovement relationships will develop, the variablity seen at the national level and in the state cases discussed in this chapter indicates that the process of relationship development needs further examination. With the state cases presented in this chapter as a starting point, the following chapters will explore that process.

"We Just Don't Work on That": Organizational Range and the Coalition Contradiction

As described previously, clean-air issues have divided unions and environmental advocates on a number of occasions. Mine workers are the ones who stand to lose the most from the strengthening of the Clean Air Act or the ratification of the Kyoto Protocol to reduce global warming. Continued reliance on coal conflicts with what many environmentalists believe to be the necessary path toward a clean-energy future. But efforts have been made to achieve some kind of compromise that would address the interests of both the mine workers and environmentalists. As noted earlier, the United Mine Workers and environmentalists were able to find common ground in 1977 when the Clean Air Act was being amended. The solution at the time was to support the use of smokestack scrubbers that would reduce the pollution released into the air as a result of coal burning. The coalition was successful at passing this legislation.

In recent years environmentalists have become more adept at factoring in the economic and job impact of environmental policies. They often seek to tailor their approach to accommodate those who might be threatened by a particular policy, including workers who fear job loss. According to David Hawkins of the Natural Resources Defense Council (NRDC), "In the political process, we often have to evaluate how [worker] issues are going to be resolved. . . . If you want to change the status quo you have to figure out a strategy that will allow you to build a majority" (personal communication, June 4, 1999).

But in some instances environmentalists feel constrained as to the extent to which they can accommodate worker concerns in the legislation they advocate, even if this may result in the failure to pass any form of the desired legislation. Hawkins and the NRDC have had contact with the United Mine Workers at various points since the 1980s over issues such as climate change and acid rain. Although it has made accommodations to mine workers on certain issues, Hawkins describes the NRDC's inability to fully address worker concerns:

To some extent, we have to prioritize the issues that we work on. . . . But if we take the position at the beginning of a political fight that says "nothing should be enacted until and unless the worker issues are addressed to the satisfaction of the workers," we are hobbling our advocacy efforts. So, we don't think we can do that. That's the case with the climate change issue. We are saying the climate change commitments need to move forward. But we didn't link it to satisfying each and every union that might be potentially adversely affected. . . . We started talking with mine workers about climate change back in the eighties. We recognized that that was an issue that was going to affect them, and we did the same thing on acid rain. We had lots of conversations with the mineworkers on that. But we were never able to reach agreement on a position. They got to the point where they were prepared to lend their support to a program, but it was a program that didn't get enough pollution out of the air. We can't do that. Our organizations aren't set up to put the union interests first, they are set up to put environmental interests first. (personal communication, June 4, 1999)

In some instances it is clear that compromises that would address workers' concerns would require too great a departure from the central goals of the environmental community for them to lend their support. But how much of a departure is "too great"? How did the NRDC come to the conclusion that the program endorsed by the United Mine Workers "didn't get enough pollution out of the air"? In some cases environmentalists fail to gain passage of any legislation regarding a particular issue because they feel that every compromise that appears politically viable is too much of a betrayal of their primary mission. This raises important questions about how each organization defines its goals and the strategic calculations that determine whether cooperation will emerge around a mutually agreeable position or whether labor-environmental conflict will erupt.

The dominant perspective among scholars studying the political behavior of social movement organizations is that groups will behave strategically to best advance their interests. Applied in the case of labor and environmental organizations, on the most basic level unions and environmentalists will get along or find themselves in conflict with one another depending on how a given policy advances or threatens their respective organizational goals. Unions primarily seek to protect the economic well-being and workplace interests of their members, and environmental advocates seek to protect the environment. When these goals clearly coalesce based on the effects of a given policy, we should expect cooperation, and when a policy yields outcomes that contradict the goals of one sector or another, we should expect conflict. But the actual behavior of these movement organizations is not so simple or obvious. As will be seen in this chapter, many factors shape the way in which a movement organization behaves. What action it will choose in a given situation cannot be inferred directly from the way in which presumed interests align. Of course, it is useful to begin with an assessment of interest alignments. The action undertaken by movement organizations can be seen as at least in part derived from assessments of these interests. But interest assessment alone does not present a complete picture because it ignores the way in which goals are constructed and the strategic assessments that go into advancing those goals. Only in the most extreme cases can we make accurate predictions about interorganizational relations based upon a crude interest assessment.

This is not to suggest that an assessment of interests among labor and environmental actors is useless. The national cases of labor-environmental conflict and cooperation and some of the evidence derived from research at the state level offer some support for the view that basic interests do shape coalition outcomes. There are some patterns in terms of how intermovement relations are influenced in the context of certain political and economic conditions. The existence of significant patterns in these relations across states suggests that the convergence or divergence of basic interests influences how movement sectors interact. And there are some obvious reasons why that may be. For example, labor-environmental relations tend to be worse where there is a significant timber industry. Conflict can be expected in areas where the timber industry is a significant force because the environmental goal of protecting forests contradicts timber industry worker interests in protecting their jobs, for which forests must be cut. A basic clash of interests yields interorganizational conflict.

Yet although certain patterns of cooperation and conflict have been identified based on political and economic conditions, the patterns are not always consistent. In other words, the tendencies toward cooperation and conflict between certain groups under certain conditions are just that, tendencies; there is still a great deal of variation in terms of intermovement relations that needs to be explained beyond the apparent political and economic interests of movement sectors. In the following chapters I will address several additional factors that shape intermovement relations, but the first factor to consider is how the general concerns of a collection of individuals, be they steelworkers or environmentally concerned citizens, take shape in the form of organizations with specific goals. An examination of the nature of such organizations will provide some insight into how they define their goals, which in turn will allow us to examine how those goals coalesce or conflict with the goals of other organizations.

The central issue that is often overlooked in simplistic strategic-interest analysis is how interests or organizational goals are defined. Was it in the interest of timber industry workers to join with the timber industry owners to fight environmental measures, or might a different strategy have yielded a preferable outcome for those workers, in terms of both job protection and other concerns that they may have had regarding their quality of life? Could they have worked with environmentalists to advance a forest preservation policy that would have protected more jobs? Could environmentalists have better advanced the environmental cause by making job safeguards a higher priority in their proposed solutions, thus winning an important ally and preventing an antienvironmental worker-industry coalition? Different strategies are conceivable in which worker and environmental interests could have been more successfully advanced. Even in a case as seemingly unambiguous as this, the basic goals of the various organizations involved do not necessarily determine that interorganizational conflict will inevitably emerge. This indicates that understanding intermovement relations goes beyond an examination of the presumed interests of the actors involved and calls for an examination of how interests and organizational goals are defined. What interests is a particular organization attempting to protect, and what specific goals does it hope to achieve? These are key issues that require further examination.

Resource mobilization theory is perhaps the most common perspective used to understand the behavior of social-movement organizations

(McCarthy and Zald 1973, 1977). According to this theory and its broader organizational progenitor, rational-systems theory, all organizations are formed in order to efficiently pursue specified goals (Scott 1992). But whom those goals are designed to serve can have a significant impact on how an organization operates, including whether or not coalition building will be part of its strategy. Organizations can be broken down into two very general categories according to their goal orientations: There are those that strive to achieve goals designed to serve the general public or some broad segment of the population (the adherents to which McCarthy and Zald [1973] refer to as "conscience constituents"), and there are those whose actions are designed to bring private gain to group members (organizations composed of "beneficiary constituents"). This public-private distinction in goal orientation is one of two very important organizational considerations necessary to better understand organizations' propensities to form or not form coalitions. The second consideration is what I term "organizational range," which refers to the extent to which organizations conceive of their missions in broad or narrow terms and whether they work on a single set of issues or a range of issue areas.

There are general tendencies that characterize both types of organizations under consideration here. Unions, at least in the United States, tend to focus specifically on the interests of their members. They seek to do so, however, on a relatively broad range of political issues, especially when we consider the work of labor federations. In contrast, environmental organizations tend to focus on public goods, yet the scope of issues on which they work is quite narrow. Most environmental organizations work exclusively on environmental issues and are rarely involved in broader social issues such as education, welfare, and housing. These differences in organizational orientation and range have important implications regarding the strategy of intermovement coalition building. There is, of course, a great deal of variability in the approach adopted by organizations from both movement sectors. This allows some to readily form coalitions, whereas others are restricted in their capacity to do so by their particular organizational tendencies. But analyzing the general inclinations of these two movement sectors in terms of the type of interests they serve and the range of goals they strive to achieve can provide some insight into the broader forces that act for and against intermovement coalition formation.

Beyond the issues of the breadth or narrowness of the goals pursued by different organizations and the interests they seek to advance, there is one goal that is common to all organizations: the goal of maintaining the organization itself (Zald and Ash 1966; Zald and McCarthy 1980). As we will see in this chapter, this goal figures heavily in determining the political strategies adopted by SMOs, including coalition building. There is disagreement among movement scholars as to whether the quest for organizational survival should be conceived of as a means to the end of achieving the stated organizational goals or whether it becomes an end in itself (Michels 1949; Rucht 1999). But what is clear is that this particular need generates contradictory tendencies among organizations in terms of the range of issues they adopt and their participation in coalition activity. In this chapter I will utilize case study evidence to analyze the way in which these conflicting tendencies can influence intermovement cooperation.

This chapter is broken down into four sections. In the first I examine the political structure that has generated the form of social movement politics that exists in the United States and the growing importance of coalition formation as a strategy. Then in the second, I consider the formation of movement organizations and the strategies that they use to maintain themselves and to advance their agendas, be they the pursuit of public or private goals. In the third section I analyze how environmental SMOs and labor unions face different constraints in regard to coalition formation depending on their organizational range. In the final section I argue that organizations in both sectors are undergoing an expansion of their ranges that will allow for increased coalition activity in the future.

The U.S. Political Structure, Social-Movement Politics, and Coalition Strategies

In considering the focus and breadth of a particular organization's goals, it is first important to analyze the origin and role of political organizations in general within the context of American democracy. It has long been recognized that in a democracy, individuals with similar interests or concerns will, in one way or another, form groups to advocate on behalf of those interests or concerns. But the form and manner in which they do so is highly contingent upon the structural environment in which they operate. According to McCarthy, Britt, and Wolfson (1991), "When people come together to pursue collective action in the context of the modern state they enter a complex and multifaceted social, political and economic environment. The elements of the environment have manifold direct and indirect consequences for people's common decisions about how to define their social change goals and how to proceed in pursuing those goals" (46).

In examining this social, political and economic environment, it is first important to note that the United States' founders had deep reservations about the influence of organized constituencies on the political process. In the *Federalist Papers*, James Madison referred to "the mischiefs of faction" and defended measures that would limit the power of constituents organized around particular interests (Madison 1961).

Given these concerns, efforts were made in the design of the U.S. government to prevent specific organized interests, be they political parties or groups outside of government, from gaining undue influence over the process of governing. In particular, the separation of powers created conditions under which political parties and, as a consequence, organized interest groups would remain weak. The absence of a unitary governing system prevents the corporatist arrangements that allow for strong and unified interest groups in other democracies (Wilson 1993).[1] With the separate branches of the U.S. government lacking the exclusive power to recognize organized interests and bestow official status upon them, the field of interest groups in the United States can be characterized as weak and fragmented.

Without a single officially recognized voice to represent a particular interest, numerous organizations may develop to advance that same cause. As a consequence each one has relatively less power in a political field populated with several voices all claiming to represent the same interest (Salisbury 1990). The multitude of environmental organizations in the United States serves as an example of this fragmentation and the weakness that results in terms of any individual organization's power.

No single organization holds very much power, nor are they capable of advancing particular goals on their own. Even the largest environmental organizations are incapable of advancing policies without the support of others both inside and outside the environmental community. This institutional weakness also plagues organized labor in the United States, as different independent unions in some cases compete with one another to represent the same workers. Even the encompassing labor organizations, such as the national AFL-CIO and the state labor federations, are weak relative to their European counterparts in that affiliation with these encompassing organizations is voluntary and subject to the whim of numerous independent international and local unions.[2]

Yet despite the institutional barriers designed by the nation's founders to limit their power and their ability to influence the political process forming organizations within a democracy to advance goals collectively is still highly advantageous; thus numerous such organizations have formed. In fact the United States stands out in terms of the number of voluntary organizations that populate the political field. As early as 1835 Tocqueville (1990) noted the significance of these organizations in the United States and the exceptional tendency of Americans to participate in such associations. Based on some estimates, as many as two-thirds of U.S. adults belong to one or more voluntary organizations (Verba and Nie 1972). Although there is some evidence that the amount and quality of civic engagement has declined in the last few decades, the number of political organizations has actually increased, and participation in these organizations is still considerable (Putnam 2000).

Although these voluntary political organizations have existed for much of America's history, until recently they played a fairly limited role in the routine creation of policy. Dramatic waves of popular protest have periodically generated significant changes in government policy; during "normal times," however, their role has been fairly restricted. Many scholars argue that for most of U.S. history, certain elite factions have dominated electoral politics and policymaking and have used their power to protect their particular interests (Domhoff 1967; Lowi 1969; Mills 1956; Ripley and Franklin 1984; Schattschneider 1960). The dominant groups identified in most cases are those composed of economic elites and the corporations, trade associations, and broader economic associations with which they are affiliated. Some have argued that the very structure of capitalist democracies creates conditions that necessitate policies favorable to these elite interests (Cohen and Rogers 1983; Poulantzas 1969; Przeworski 1985). Others have focused on the way in which elite actors are directly involved in exercising political control through such means as financially sponsoring political campaigns, restricting the political nomination process, and holding political office (Domhoff 1967; Miliband 1983; Mills 1956).

Not only are moneyed constituents able to control the electoral process, but through the internal workings of government decision making, public policy is shaped to reflect their interests. The close and closed relationships between these private-interest groups, elected officials, and public-agency administrators have been referred to as "subgovernments" (Ripley and Franklin 1984) and "iron triangles" (Gais, Peterson, and Walker 1984; Salisbury 1990). These discreet relationships enabled private actors to "capture" state agencies. Individuals sympathetic to the interests of the relevant elites are appointed to administer government agencies, where they proceed to make and enforce policies in ways that benefit their supporters without extensive public oversight and contrary to the stated mission of the governmental agencies involved.

Political changes that have taken place since the 1960s, however, have led most scholars to modify their view of the current process of policy development and implementation (Davis and Davis 1988; Gais, Peterson, and Walker 1984; Jenkins-Smith and Sabatier 1994; Loomis 1986; Sabatier 1988; Sabatier, Loomis, and McCarthy 1995; Salisbury 1990). Two factors in particular have led to this reevaluation of policy formation. First, since the 1960s there has been explosive growth in the number and importance of organizations focused on issues of government policy (Almond and Verba 1972; Berry 1984; Salisbury 1990; Schlozman and Tierney 1986; Walker 1983; Wilson 1990, 1993). In 1955 there were roughly 5,000 national associations. Today there are over 23,000 (Hula 1999, 3). A large percentage of the multitude of organizations focused on issues such as the environment, public health, and consumer protection were founded during this period. Looking at the environmental movement alone, what began as a handful of environmental organizations in the early 1960s grew in number to as many as

10,000 by 1990 (Putnam 2000, 155). Second, government programs have expanded over the same period and government policy has become increasingly complex (Walker 1983). The creation of regulatory agencies such as OSHA and the EPA as well as a significant expansion of welfare and social-service programs, along with the agencies responsible for overseeing them, has taken place since 1970. These government agencies have created a host of new rules and enforcement mechanisms that serve as the focus of new organizations concerned with issues addressed in the expanding public policy universe.

Although these two factors are related to one another, it is difficult to determine precisely the direction of causality. Some have noted the role of new organized movements in generating and expanding government programs, whereas others have pointed to the way in which government actors in new public-policy arenas encourage and facilitate the development of organized constituencies to defend and promote the programs that they administer. Resource mobilization theory would suggest that expanding resource pools developed in the prosperous years following World War II allowed for organizational proliferation (McCarthy and Zald 1977). Citizens who found themselves with more disposable income became the target of appeals by movement entrepreneurs seeking to amass resources that could be used to shape public policy.

Several other factors that may have contributed to the social-movement explosion of the 1960s include the higher levels of education achieved during this period, the development of inexpensive mass communications, and the proliferation of organizational experience and knowledge that was spread through the Civil Rights and student movements of the 1950s and 1960s (Berry 1984; Walker 1983). Open-meetings laws, a weakened congressional seniority system, and the increasingly important role of legislative subcommittees and staff members in public-policy formation also served to decentralize power and made the governing system more accessible to outsiders. Regardless of the impetus for the growth of the social movement sector as a whole, the new interest organizations that burst onto the scene during this period disrupted the relatively insulated policymaking relationships that had previously existed by increasing scrutiny of the policymaking process and providing the organizational vehicles necessary to challenge policies on the basis of the public interest. Elite groups, multinational corporations in particular, still clearly retain a disproportionate share of political power and influence in the United States. Economic globalization has arguably increased their power in recent years. Current political arrangements are still far from the egalitarian pluralist ideal touted by some, but the growth of popular-movement organizations in recent decades has altered the terrain on which the political struggle is carried out.

The proliferation of social movement organizations in the United States has also occurred at a time when the significance of the traditional form of popular political-interest representation, political parties, has declined. It has already been noted that the nonunitary system of government in the United States was designed to disperse power and prevent factions from gaining undue influence. Although not perfectly effective in achieving this goal, it did leave political parties relatively weak, and in recent decades they have become even more impotent (Crotty 1984; Greider 1992; Wattenberg 1984).

The political-party system in the United States has been characterized as "candidate driven." Candidates for office at any level are not entirely dependent on a political party, and there are few mechanisms that can ensure accountability to the party platform (Collie 1984). Incentives can be offered to encourage adherence to the party's positions in the form of campaign funding and the award of appointments to legislative committees for loyalty to the party's platform or leaders. If a candidate is capable of raising their own campaign funds, however, and they are successful in their bid for election, there is little that party officials can do to make them adhere to partisan dictates once in office. A good example of this occurred during the 2000 debate about renewing most-favored-nation trading status to China. The Clinton administration pressed Democratic members of Congress hard on this issue, yet many still opposed granting such status, largely because of pressure from labor unions, environmentalists, and human-rights groups. It was organized movement groups, not party affiliation, that determined the position of many lawmakers on this issue.

In addition to this ongoing institutional weakness, the significance of political parties has eroded further in recent years. As measured by party identification, the growth in issue-based voting, and the public confidence expressed in them, political parties are in decline (Berry 1984; Broder 1978). These factors, especially the growth in issue-based voting, point to the relatively increasing importance of other forms of organized interests in both electoral politics and policy development. It is these movement organizations, more so than political parties, that now mobilize voters and resources to advance political objectives. Recent elections offer further evidence of this, as the amount of funding dedicated to "issue education" by social movement organizations increased dramatically in these elections. Issue education is a strategy in which SMOs seek to educate the public directly about the candidates' positions on certain issues, independent of any coordination with the candidates' political parties. Elected officials are increasingly held more accountable to these independent movement organizations than to the political party with which they are affiliated. Changes in campaign finance law may influence the means by which political parties and independent political organizations act to advance the issues of concern to them, but there is no question that a broad range of movement organizations and interest groups now exercise considerably more power than in the past.

It is clear that social movement organizations now play a very significant role in the shaping of public policy. The irony of the proliferation of these groups, however, is that although they have gained in importance overall, individually they are often too weak to achieve their goals. The structure of political institutions in the United States has given rise to numerous organizations while at the same time curtailing their individual power (Salisbury 1990). For example, historically, the AFL-CIO has had considerable influence over the Democratic Party (and still does, albeit to a lesser degree). But the Democratic Party must now consider policies being advocated by the federation in light of the interests of environmentalists, women's organizations, senior-citizen groups, and numerous other organized constituencies on whom Democratic officeholders must now rely for support. Most of the time, the goals of these popularly based organizations do not conflict with one another. But in instances in which there are disagreements, different constituents can be played off against one another. Officeholders are in a position to choose among them, and given the number of other groups that they can count on for support, each individual group has relatively little power.[3]

But it is not just conflict among newly mobilized constituencies that inhibits the individual effectiveness of each group. The large number of organizations populating the political arena itself renders individual organizations less capable of determining policy outcomes. Officeholders must still respond to the pressure of organized constituents; the growth in the number of interest organizations, however, has necessitated changes in such organizations' strategy for achieving their goals. The role that political parties formerly played in terms of synthesizing the interests of diverse constituents has declined as the significance of parties themselves has waned. What we are left with is a disorganized political field in which numerous organized interests attempt to exert pressure on individual representatives in order to advance their particular cause. This combination of weakened parties and multiple competing interest organizations has necessitated that these organizations attempt to coordinate their efforts to make their voices heard. This is where the importance of coalitions becomes apparent. The AFL-CIO alone may be incapable of advancing a particular policy goal, but if other groups can be brought on board to rally around the same cause, together they may have the critical mass necessary to advance the policy goal.

In recent years numerous scholars have noted the increasing importance of coalition activity in the legislative and policymaking process (Berry 1984; Brecher and Costello 1990; Bystydzienski and Schacht 2001; Hojnacki 1997; Hula 1999; Jenkins-Smith and Sabatier 1994; Keller 1982; Loomis 1986; Moe 1980; Sabatier 1988; Salisbury 1990; Schlozman and Tierney 1986; Wilson 1993). This growing importance of coalitions is evidenced in two ways. The first factor indicating the increasing importance of coalitions is the growth in their numbers and the frequency with which organizations now join in coalitions. Ninety percent of organizations report joining in coalitions and doing so more frequently than in the past (Schlozman and Tierney 1986). The second factor is the growth of larger and larger formal coalition organizations. To use private business organizations as an example, we have witnessed a progression from individual-firm advocacy to trade associations to superassociations such as business roundtables and the chambers of commerce to broad coalitions that include organizations from outside the business community (Keller 1982).[4] The coalitions of interest in this study those that involve labor unions and environmentalists, are analogous to this highest level of coalition activity. Whereas like-minded organizations focusing on the same set of issues, such as environmental groups, often work together, coalitions involving disparate groups such as environmentalists and unions are more rare. Nonetheless, the trend indicates that this type of coalition will become more common. Given the strength of this trend, the central question is what type of coalition configuration will emerge as dominant or typical and whether unions and environmentalists will be aligned or in opposing camps.

In addition to the pervasiveness of coalitions, there is also strong evidence to show that the support of others is very important to the success of organizations in achieving their goals (Brecher and Costello 1990; Bronfenbrenner et al. 1998; Clawson and Clawson 1999; Steedly and Foley 1990). Using data gathered by William Gamson (1990) in what is probably the most comprehensive assessment of the effectiveness of social movement strategy, Steedly and Foley (1990) determined that the support of other organizations is the second-most-important variable in determining successful movement outcomes. Recognition of the importance of interorganizational support has led national labor leaders to advocate more community outreach and coalition building.

Clearly in the United States today, corporate interests and other elites have disproportionate influence on government policymaking. But popular mobilization in recent decades has given rise to a multitude of organizations that challenge elite control of the policymaking process. Relatively speaking, policy today is not so much the result of pressure from one or a few actors with a concentrated interest in a policy area as it is the result of competition among a wide range of organized interests. Popularly based interests, though disadvantaged in terms of the resources that they individually control, can score important political victories if they are able to pool resources and coordinate activities. Thus coalition formation is a crucial political strategy for popularly based organizations. But the nature of such organizations creates some contradictory tendencies that can inhibit them from adopting the very strategy that would help them advance their goals. A closer look at these organizational tendencies is necessary to see why unions and environmentalists sometimes find it difficult to unite.

Mobilizing Resources and Members

Movement organizations now play a very important role in the creation of government policy, and they exercise influence in many cases by forming coalitions with one another. Environmental organizations and labor unions are among the most significant of these popularly based movement organizations, and when they work together, the resulting labor-environmental coalitions have proven to be very effective (Dreiling and Robertson 1998; Obach 1999; Rose 2000). Although coalition formation can be a good strategy for advancing organizational goals, movement leaders must also concern themselves with maintaining their own organizations and making sure that the central mission of their groups remain in focus (Zald and McCarthy 1980). Thus, to understand when organizations are free to enter into coalitions and when they are not, we must first identify how the organizations themselves come into being, whose interests they are designed to serve, and how they determine their organizational range.

Early social movement theorists argued that movement activity and the organizations associated with it arose as a result of social or economic disturbances (Truman 1951). Such disturbances were thought to negatively affect certain individuals who, in response to their grievances, would then organize themselves into groups to advocate collectively on their own behalf. These theorists left unexamined, however, the actual process through which mobilization and organization takes place. Critics later argued that mobilization is not spontaneous and that in fact there are significant barriers to individual participation in movement activity even when grievances are widespread.

In *The Logic of Collective Action*, Mancur Olson (1965) employed a rational-actor framework to challenge conventional ideas regarding individual participation in collective action. He argued that given the nature of collective goods, a rational self-interested individual would not participate in efforts directed at attaining such goods. Due to the lack of limitations on the distribution of collective goods, any gains made through the action of a group are available to all, regardless of their participation in the group's action. Thus, an individual worker might say to herself, "Why go out on strike and lose work time and wages and possibly even my job? One more person on the picket line will not make any difference to the outcome, and if the rest of the striking workers are successful, I'll get the wage increase along with everyone else." Similarly, an individual concerned with the health effects of ozone depletion might choose not to contribute money to an environmental group seeking to arrest ozone depletion, knowing that once the ozone depletion problem is solved, he will reap the health benefits regardless of whether he has contributed the money. What's more, if the group is unsuccessful, then the rational nonparticipant will have avoided wasting time, effort, and resources on a futile effort. According to this perspective, rational individuals will not assume the risk or expense of participation in collective efforts but rather will wait to collect the benefits won through the work of others. In this light, movement organizations face a serious, if not insurmountable, challenge when attempting to recruit members.

Although this reasoning would tend to preclude *any* collective action at all, Olson cited two factors that may lead to group participation despite this tendency. The first is the provision of "selective incentives" to those who join a particular group or participate in a particular mobilization. Goods available only to those who participate in a given effort make involvement in that effort economically rational. Like any purchase in the market, the goods and services provided directly by the movement organization may be considered worth the costs of participation. For example an environmental group may offer magazine subscriptions or access to group events that may make membership in the organization worthwhile, regardless of whether the group achieves any of its stated political goals. The second factor cited by Olson is the existence of a coercive mechanism that essentially requires that individuals participate regardless of their expected returns. Olson points to labor unions operating in a closed-shop environment as an example of such a coercive mechanism, since workers are required to contribute to the organization as a condition of retaining their employment. Thus, according to Olson, to recruit and retain participants, movement organizers have to provide a benefit reserved only for participants, or participants must be in some way forced to join the group.

Although Olson presented an important challenge to the earlier theorists who left unexamined the decisions faced by individuals who make up any movement organization, later scholars identified the shortcomings of Olson's rational-actor model and his emphasis on material incentives. A more encompassing rational framework looks beyond the purely economic considerations emphasized by Olson to include other types of incentives (Clark and Wilson 1961; Ferree and Miller 1985; Gamson 1990; Klandermans and Oegema 1987; Salisbury 1969; Tilly 1978). In addition to material incentives, some have specified two other factors that can induce participation: solidary and purposive incentives (Salisbury 1969).

Solidary incentives are those benefits that accrue to individuals through working with others toward a common goal. Friendships, camaraderie, and companionship are important benefits that may draw individuals to group activities or keep them there once they have encountered the group. Solidarity has played a central role in the formation and maintenance of labor unions throughout history. Strong mutual bonds among workers have sustained collective action even in the face of violent repression.

Purposive incentives, sometimes referred to as "expressive" incentives, refer to subjective feelings that individuals take away from the experience of contributing to a cause in which they believe. For example, those who believe in wilderness preservation receive a kind of cognitive reward when they actually act to advance that principle. Participation of this kind allows individuals to express who they are in a way consistent with their beliefs and self-image. Although still confined within a purely rational-actor framework, this expanded interest model gives appropriate consideration to nonmaterial factors that determine individual participation in group efforts.

Still others have rejected the individualist orientation of private-interest theory (Gamson and Fireman 1979; Muller and Opp 1986). Muller and Opp, for example, propose a "collective rationality" in which participants in a movement do not operate on the basis of individual incentives but rather act in ways that are rational when undertaken by a collectivity. Although objectively each individual is only making the decision to participate themselves, that decision is rational if it is evaluated from the perspective of an entire group: If collective action is the only means to achieve a particular goal, then it is rational for individuals who desire to achieve that goal to act collectively. The fact that each individual can control only their own behavior and not that of the entire group does not prevent them from identifying and acting on the logical connection between group action and goal achievement.

The question of what motivates individuals to participate in collective action is not only of interest to movement theorists; it is also of crucial importance to organizational leaders and founders. In a crude, but conceptually useful, comparison to business enterprises, some refer to social-movement leaders as "political entrepreneurs" (Salisbury 1969). Like business executives, who must ensure that the firm is profitable, political entrepreneurs must run the organization, devise ways to attract members, and ensure that the group is maintained through their own assessment of effective incentives. In *The Organization of Interests*, Terry Moe (1980) describes the role of the political entrepreneur:

the political entrepreneur's situation is analogous to that of the economic entrepreneur in simple theories of the business firm. In broad outline, the requirements of survival are the same in both cases. The entrepreneur can be viewed as investing capital in a set of benefits (collective goods and selective incentives) that he offers to a market of customers (potential members) at a price—the cost of joining the group (dues) plus any fees that may be attached to any particular services. Potential members buy the offered package of benefits, and thus join the group, only if the value of the benefits is expected to be greater than their price; and they will remain in the group as long as this continues to hold true. (37–38)

The question then becomes, what package of incentives must organizational leaders offer in order to mobilize the members and resources necessary to advance the organization's cause and maintain the organization?[5] This is where the distinction between public and private benefits becomes important. Is the stated purpose of the group to advance some public good, or is the group designed to achieve benefits for the individual members who make up the group? This distinction will significantly shape the choices made by group leaders regarding which kinds of incentives to offer to attract members. From there we can consider how the need to motivate participation can shape the organization's political strategy and relations with other movement organizations.

Organizations that are centered on public-goods provisions face different contingencies in their recruitment strategies than those that are oriented around providing private material goods (Jenkins-Smith 1991;

Moe 1980; Sabatier and McLaughlin 1990; Scott 1992; Wilson 1973; Zald and Ash 1966; Zald and McCarthy 1980). Essentially, those organized around public, nonmaterial goods are going to have to rely more heavily on purposive incentives. Selective material and solidary incentives will still play some role in recruiting members for such organizations, but an organization based on the attainment of goods that will not be restricted to members must place relatively more emphasis on the purposive element. Since members are receiving little or no private material reward, it is the sense that they are contributing to and advancing a particular cause in which they believe that is likely to motivate members to participate in this type of movement. This is true regardless of whether one conceives of potential recruits as operating on the basis of individual rationality or on the basis of some sense of collective rationality. The task of the recruiter is the same, motivating individuals to act on their principles by supporting the organization. In contrast, those organizations built primarily around achieving private material goals will have to rely more upon material incentives to satisfy their members.

Thus far, we have considered recruitment for movement organizations based primarily on the rational choices of individuals given a set of incentives offered by those organizations. Resource mobilization theory adopts this approach to social movement study. Organizations are the focus, and organizational leaders strategically calculate how best to attract participants (McCarthy and Zald 1977). Yet others argue that the rational-actor approach, although suitable for assessing participation in some early social movements, does not apply well to recent mobilizations. Some have conceptualized this distinction in terms of "old" and "new" social movements (Larana, Johnston, and Gusfield 1994; Melucci 1980; Offe 1987). New Social Movement theorists point out that traditionally, most movements have been class based and oriented toward economic goals. The labor movement is the ideal typical manifestation of such a movement. In contrast, many of the movements that developed in the 1960s are neither class based nor focused on material issues. Movements in support of peace, gay rights, the environment, and women's liberation are counted among these new social movements. New Social Movement theorists argue that these movements tend to embody "postmaterial values" (Opp 1990) and are motivated not by the quest for material gain, but by a sense of identity or cultural attributes that correspond with certain forms of movement participation (Downey 1986; Epstein 1991; Larana, Johnston, and Gusfield 1994; Taylor and Whittier 1992). It is thought not that people participate in new social movements because they hope to gain materially or in any other concrete way, but rather that they express and reinforce their identity through their participation in such movements.

Based on this perspective, participation in labor unions, which represent an old social movement, is best understood as being motivated by the pursuit of material goods. Through unions workers were able to increase wages and to secure other direct personal benefits on the job. In contrast, participation in new social movements, including the environmental movement, is thought to be inspired by less rational motives associated with cultural aspects of the individual's social milieu and the sense of identity that they posses or seek, which is associated with certain forms of political behavior. If one thinks of oneself as someone who is concerned about the environment, then joining an environmental organization is a way to express that identity. Membership then reinforces that sense of oneself as an environmentalist. Unlike that involved in being a member of a labor union, the potential for direct personal gain resulting from being part of an environmental organization is remote. New Social Movement theorists argue that it is clearly something other than the rational pursuit of personal gain that underlies participation in movements of this type.

Critics of the New Social Movement approach have charged that the distinctions New Social Movement theory makes between new and old social movements, and individuals' motivations for participating in them, are overstated. For example, some have shown that the same cultural or identity-driven behaviors associated with new social movements can be found to motivate actors within the labor movement (Fantasia 1988). Again, as Olson pointed out, even individual participation in collective action from which one stands to benefit directly is not necessarily rational; thus union members do not represent purely self-interested rational actors. A sense of identity tied to being a union member serves the same function that New Social Movement theorists tend to associate exclusively with the newer social movements. It is also possible to conceptualize the identity function emphasized by New Social Movement theorists as an incentive parallel to the expressive or purposive goals that rational-choice theorists cite. In this light, the motivational processes involved in participation in new and old social movements are not so distinct from one another.

Regardless of how these factors are conceptualized, debates regarding the different motivations for participating in social movements provide insight into some basic organizational differences between unions and environmental organizations. These two types of organizations do have very different goals that serve as the main motivation for their activity, and a basic distinction can be made between those centered primarily on the promise of personal return to members and those whose goals are designed to bring benefit to the broader society. Despite the incidental benefits that may accrue to the broader community as a result of their presence, in recent history within the United States, labor unions have focused primarily on attaining private material benefits for their members. This presents a contrast with the nonmaterial, public-goods objectives of some organizations considered to be among the new social movements, including the environmental movement. This may or may not mean that there are differences in the self-concept and motivations of the participants in the labor and environmental movements.

The issue of importance here, however, is not so much what actually motivates individual participation as it is what *organizational leaders perceive to be important* in terms of attracting and motivating the organization's members and maintaining the organization. Coalition formation is the central concern in this discussion, and organizations, not their individual members, form coalitions. But the motivations of individual members, or at least the perception of those motivations on the part of movement leaders, are crucially important for understanding coalition propensities. Movement leaders are faced with having to make decisions regarding what kinds of incentives are required to attract and retain members and what limitations the necessity of offering those incentives places upon the organization's political strategy. In this sense the rational, organization-focused approach of resource mobilization theory is most useful here (Zald and McCarthy 1980). Organizational leaders strategically calculate how to best advance their organization's goals—

both its stated goals and the goal of maintaining the organization itself. What incentives must be offered is contingent on the type of goals the organization is striving to achieve.

The perceived differences in the reasons for member participation from organization to organization entail differences in the recruitment and maintenance strategies adopted by organizational leaders. To revert to Salisbury's terminology, the leaders of organizations that are focused on public, nonmaterial goods must appeal more to the purposive goals of the members. New Social Movement theorists characterize this type of appeal in terms of the need to sustain the culture or foster a collective identity that is associated with movement participation. But regardless, the strategies adopted by these leaders will differ, at least in degree, from those adopted by organizational actors involved in more materially oriented movements. Although union leaders may appeal to class identities or make efforts to ensure that union membership is an ingrained element of workers' culture, they have to pay significant attention to the actual material promise associated with union membership if they wish to keep members. To maintain the union (and their own position within it), it is necessary for union leaders to advance the interests of their members, be it by increasing their wages, improving their working conditions, providing them with low-interest credit, saving their jobs, or providing some other personal benefit. Environmental organizations, in contrast, will use various selective material incentives such as newsletters or access to organizational events, but these will not provide the primary motivation for membership. Instead, member appeals will focus more on the good that one is doing and the important role that members are playing in protecting the natural environment (Zald and Ash 1966).

Of course, the actors involved in the environmental movement are quite diverse. Not all environmental organizations are oriented toward the same public goods, and some are better situated than others to appeal to the individual interests of potential members. Some environmental organizations, for example, focus on immediate threats to particular groups of people, providing these organizations with the opportunity to appeal to specific member interests. These groups are sometimes referred to in a somewhat derogatory manner as "NIMBY (not in my back yard) organizations," because their primary focus is to prevent some undesirable environmental condition in their own community while expressing little concern regarding any larger environmental threat to people elsewhere. In reality, despite the narrowness of their primary focus, such groups often incorporate broader environmental sensibilities into their local struggle, if only rhetorically. But these groups tend to be relatively small, local grassroots organizations that mobilize on a temporary basis. Because the focus in this discussion is on long-term labor-environmental coalition building, this type of organization is not given substantial consideration here.

The discussion here focuses instead on professional state or national environmental organizations. In some cases these groups may take up a local issue of particular interest to members in a specific area, but for the most part they address broader environmental issues with ramifications for the society as a whole. To some extent global or regional environmental problems can be cast as a threat to particular individuals (in terms of their health, for example), thus eliciting support for organizations dealing with those problems based on the specific interest that potential members may have in solving them. But for the reasons elucidated by Olson, this interest, by itself, will not generate widespread support for an organization or movement. The benefits received in the event of movement or organizational success will be available to all, regardless of whether they have contributed to the cause, and an individual contribution to a mass movement is meaningless in terms of determining movement success or failure. Given the impossibility of limiting the benefits of environmental protection to members of environmental organizations alone, these organizations have to rely almost exclusively upon purposive incentives to attract and retain members. In most cases these organizations will solicit membership and support on the basis of moral imperatives, and support will come as a result of the potential members' sense of "wanting to be part of the solution" as opposed to some hope of personal benefit.

The Coalition Contradiction and Organizational Range

As described above, social-movement theorists have approached the question of member recruitment and mobilization in different ways.

Some see movement participation as the quasi-rational pursuit of individual interest, whereas others argue that movement participation reflects an effort to express one's identity. There is some validity to both conceptions and a fair amount of overlap between them. But whether one conceives of the motivations for movement participation in terms of "new" verses "old" movements, "postmaterial values" versus "material interest," or "selective" material incentives versus "purposive" ones, there is broad recognition that there are some basic differences in the nature of goals that different types of social movements are striving to achieve. This is ultimately related to the central concern in this discussion regarding coalition participation. Thus far we have established that (1) the stated purposes of organizations that make up the labor and environmental movements differ in fundamental ways, (2) participants are, in turn, motivated to join such organizations for different reasons, and (3) organizational leaders must be attuned to the different incentives that motivate participation in order to maintain their organizations. This leads us to a fourth point that logically flows from these conditions: The need to provide the necessary incentives for membership can inhibit participation in coalitions as a strategy.

Environmental organizations are seeking to achieve goods that will benefit wide segments of the public, whereas unions are primarily oriented toward securing the interests of their membership. Those who attempt to build these two types of movements must strategically decide how to advance their political goals while sustaining their organizations. Organizational maintenance requires that these two types of organizations appeal to their members in different ways. The question of concern here is whether organizations based on purposive appeals can align politically with those that are based primarily on material incentives. To understand the effect that this distinction has on coalition participation, we must analyze it not just in relation to the public-private distinction but also in relation to the related issue of organizational range, the scope of the issues that an organization addresses. Depending on whether an organization is focused on public goods (as are environmental SMOs) or private goods (as are labor unions), its organizational range and coalition capabilities will be influenced in different ways. Each of these issues is examined separately below.

Environmental SMOs

For environmental SMOs (and all organizations centered on promoting public goods) the question of organizational range can be explored by first considering a key dilemma faced by leaders of such organizations. Given the *general* appeal to purposive environmental concerns that environmental organizations must make, how do they attract support for their *particular* group? As already discussed, the political structure in the United States tends to foster multiple social movement organizations. Since none can be given official status and claim exclusive representation of a given cause or interest, these organizations are left to compete with one another for voluntary members. Given the crowded field, in order to attract supporters and funds, organizations must distinguish themselves in some way in terms of what they actually do. They may seek to do this simply through the use of advertising techniques. Direct-mail marketing campaigns have become increasingly sophisticated, and those who find themselves on the right (or wrong) list can be inundated with slick promotional material and membership appeals from an array of like-minded organizations.

But despite these Madison Avenue techniques, or perhaps because of them, organizations must distinguish themselves from others to attract support. Given a limited adherent pool, "SMOs must compete . . . for the time, effort, loyalty, and money which citizens can give or withhold" (Zald and McCarthy 1980, 4). They do so in part by focusing on a particular issue or set of issues. Again using terminology drawn from the business world, Zald and McCarthy refer to this as "product differentiation."

To some extent government policymaking mechanisms necessitate specialization among organizations by creating separate agencies to address distinct policy issues. This channels movement organizations into narrow, specialized fields. Whereas at an earlier point in history, a workplace and community health hazard would have been viewed as a single issue, the establishment of the EPA and OSHA makes it necessary to have two separate spheres of expertise to deal with the two agencies on what is essentially the same issue, facilitating the creation of more narrowly focused organizations. Yet even within the socially constructed policy realms that result from this governmental structure, organizations are compelled to specialize. An organization's focus can be as broad as "environmental issues," but oftentimes having such a broad focus is not enough in itself to draw support to an organization, given the large number of organizations that address environmental protection. It may be possible for well-established groups to maintain a very general focus; upstarts, however, must be able to distinguish themselves from other groups already in existence. New groups, and to a lesser degree existing ones, are always on a quest for the "new niche market" in order to recruit supporters (Putnam 2000, 157).

Some movement scholars have recognized this challenge faced by movement organizations. Over twenty years ago McCarthy and Zald (1977) hypothesized that intensified competition among SMOs would lead professional movement organizations to focus their agendas more and more narrowly, and the growing number of issue-specific organizations offers evidence of the validity of that prediction. Even new grass-roots organizations must justify their existence. If another organization already exists that deals with essentially the same concerns, the founding members of the new organization have to be able to explain why they are starting a new group. This need for differentiation among organizations within a particular movement requires a closer examination of how organizations appeal to members.

The process of constructing appeals to members and prospective members of an organization has been examined in terms of "frame alignment" (Snow and Benford 1988; Snow et al. 1986). David Snow and his colleagues (1986) define this process as "the linkage of individual and SMO interpretive orientations, such that some set of individual interests, values and beliefs and SMO activities, goals and ideology are congruent and complementary" (464). Movement organizations must present issues in a way that either taps into the existing concerns of potential supporters or generates new sentiments that will inspire recruits to support the cause. Given the crowded field of SMOs, organizational frames must center on increasingly narrowly defined issues. Environmental SMOs may choose to focus on a particular set of environmental issues, such as clean water, rain forest protection, or saving the manatee from extinction.

The way an issue is framed also necessitates appeals regarding organizational efficacy. An organization may attempt to distinguish itself on the grounds that it is the oldest or the biggest or the most influential organization of its type. Tactical differentiation may be even more common than drawing distinctions among organizations based on the actual issues addressed (Zald and McCarthy 1980). Sometimes organizations will highlight their efficacy through the superiority of the tactics they use to advance their cause, such as their ability to generate letters to elected leaders or the expertise they bring to an issue by employing scientists who specialize in the study of a particular environmental phenomenon. By engaging in this kind of product differentiation, organizations are better able to distinguish themselves from others and in that way increase the effectiveness of their appeal to potential funders and supporters who are presented with a wide array of membership solicitations. But because these appeals are used to distinguish the organization and attract members, they can also limit the organization in terms of the strategies it uses to advance the cause it represents. Because of their very narrowness, organizations limit their capacity to join in coalition with other organizations, which, given the crowded political field and the weakness among individual organizations it spawns, is often a necessary strategy to achieve organizational goals.

The leaders of voluntary movement organizations can be considered to be in a contradictory situation in terms of the need for organizational maintenance and that for organizational effectiveness. I refer to the dissonant pulls toward these two needs as the "coalition contradiction." As noted earlier, organizational leaders have not one but two key objectives: to advance their cause and to maintain their organization. As previously discussed, the emphasis on organizational maintenance should not be overstated, nor should it be interpreted as merely a vehicle for the careers of those employed within it. Organizational leaders and staff members tend to be very committed to the political causes on which they work (Sabatier and McLaughlin 1990). Maintaining their organizations is best understood primarily as an instrumental effort to advance the cause for which the organization was created, not an end in itself. Nonetheless, maintaining the organization is very important for the purpose of achieving the organization's goals. Even when an environmental policy victory is realized, rarely can any organization claim that its goals have been fully achieved and that thus there is no more need for the organization.

Granted, those employed by such a group have an obvious personal interest in maintaining the organization, but even without that vested interest, anyone concerned about the environment would agree that there is no benefit to closing the doors of an effective environmental organization. There is certainly no shortage of environmental issues to be addressed. Environmental organizations may or may not be able to make the transition to another cause, once they have achieved the specific goal for which they were formed, but it is likely that additional concerns beyond the initial focus of an organization will inspire some to attempt to sustain it after the initial focus is no longer applicable (Gamson 1990). In any case, leaders who are concerned about protecting the natural environment and view their organization as a means for achieving that end must make efforts to maintain the organization in an ongoing way.

Coalition activity poses a dilemma in regard to the dual goals of maintaining the organization and effectively advancing its stated goals, because coalition activity often necessitates some modification of the organization's policy goals so that they better correspond with the goals of other organizations. Although coalitions are often necessary to advance a political cause, veering too far away from the specific policies favored by an organization (those initially used to attract members) in order to make the organization a viable coalition partner can threaten the organization's viability and thus lessen its prospects for achieving its original goals. Thus an organization's goals must be distinctive to attract members and maintain the organization, yet they must still have enough commonality with those of other organizations to enable the organizations to work together.

Hathaway and Meyer (1997) examine this issue in terms of two variables, political agreement and market overlap. According to their model, organizations are more likely to work together the more similar their political goals, unless their target supporters overlap, in which case they will instead compete with one another. Hathaway and Meyer hypothesize that organizations with similar goals, but complementary strategies and separate bases of support, are most likely to work together. For example, professional lobbying organizations would be more likely, according to Hathaway and Meyer's hypothesis, to work with grassroots direct-action organizations than with other lobbying organizations because, as professional lobbyists working on similar issues, they are likely to be in competition with one another for membership. The direct-action group, though addressing the same issues as the lobbying group, appeals to a different membership base; thus there is no competition for scarce resources (in this case, member participation and support).

This is an interesting proposal; most empirical evidence, however, does not bear it out. Suzanne Staggenborg (1986), in her analysis of prochoice organizations, found that those with similar strategies are more likely to work together. The common approach used by similar organizations allows much more opportunity for coordinated action. For example, professional lobbying organizations are more likely to work with other professional lobbying organizations than with grassroots groups. This can be seen in the relatively high amount of coordination among professional environmental organizations based in Washington, D.C. The major Washington environmental organizations hold monthly meetings to share ideas and coordinate political strategy despite significant overlap in their goals and targeted membership base. Shared political goals and a desire to achieve them through coalition work tend to override competition between organizations for members.

When coalition activity will be undertaken depends in part on the extent to which such a strategy allows an organization to remain focused on its stated goals versus the extent to which it requires some deviation from those goals. In some cases the goals of distinct organizations will converge in a single policy, allowing all of the organizations to work cooperatively without any modification of their fundamental objectives. In rare instances organizations may join with others in an effort to achieve goals that were not originally included in their organizational missions, despite the risks this poses to organizational maintenance. In still other instances organizational goals may be partially adjusted to satisfy coalition partners, allowing each to achieve a partial victory. To better distinguish the types of coalition activity in which organizations are likely to engage, forms of intermovement cooperation have been broken down into three categories: instrumental cooperation, compromise cooperation, and enlightened cooperation (Obach 1999). I will first describe these three forms then return to the question of organizational range and which groups are most capable of and most likely to engage in each of these different forms of intermovement cooperative action.

We saw in an earlier chapter, when examining the effect of various political and economic factors on labor-environmental relations, how the pursuit of an organization's core goals can, in some cases, be directly enhanced through coalition activity. In instances in which the goals of distinct organizations converge in some way, a strategy of coalition formation may provide the opportunity for the benefits of coalition efficacy without any of the individual organizational drawbacks associated with the modification of goals. This is referred to as "instrumental cooperation."

For example, the Sierra Club in the state of Washington worked collaboratively with the union representing tugboat workers to gain passage of legislation requiring tugboat escorts for oil tankers moving through the Strait of Juan de Fuca off Washington's Olympic Peninsula. Noting the devastating effect of the 1989 *Exxon Valdez* oil spill in Alaska, environmentalists saw the threat that a major spill in Washington's waterway posed to the surrounding environment. Environmentalists viewed tugboat escorts as one way to lessen the environmental risk, and tugboat workers saw the obvious benefit of more jobs should the requirement for tugboat escorts be imposed. Numerous other examples of this type of cooperation exist. Chemical workers have collaborated with environmentalists on the control of toxic substances that pose a threat to worker health and the environment. Steelworkers have worked with environmentalists based on the job creation potential of solar energy and air pollution measures. Instrumental cooperation is easy to understand, and it represents the most common motivation for cooperative action on the part of distinct movement organizations. The majority of research on coalition behavior has this kind of cooperation as its focus. This type of cooperation can also pave the way for other types of coalition activity (Nissen 1995).

In some cases, for cooperation to occur, partners in a coalition may be required to diverge from their stated agendas. Coalition work between two organizations may be possible, for example, only if both organizations compromise somewhat on their goals in order to accommodate the concerns of their potential partner. This is referred to as "compromise cooperation." In essence, organizations are sacrificing some of what they want to increase the probability that they will achieve part of what they want. This probability of partial success is increased by virtue of the fact that partners in the coalition are now supporting the cause along with the organization.

A case cited earlier from Wisconsin in the 1980s demonstrates the pattern typical of compromise cooperation. When unions and environmentalists in Wisconsin clashed over a bill supported by environmentalists that required that deposits be placed on beverage containers, a compromise was achieved in the form of a recycling program. Both sides sacrificed some of what they hoped to achieve, but through this action, they were better able to secure a partial victory. According to Ken Germanson, a staff member of the Allied Industrial Workers and the coordinator for the Wisconsin Labor-Environmental Network,

One of the key areas of concern between labor and environmentalists were so called "bottle bills" and can legislation. At that time we still had several thousand workers in the brewing industry in the state, which were all good union paying jobs, plus we had a lot of workers in the can industry. And the unionists didn't care for deposit legislation on cans and bottles because it would, they felt, do away with jobs. . . . The environmentalists of course wanted deposit legislation. . . . Eventually we were able to jointly support the recycling legislation. [The environmentalists] knew that deposit legislation was going to be dead on arrival in the legislature, because we in labor had such strong support among the legislators. . . . Deposit legislation was going to be doomed. But . . . we came up with a compromise position [on recycling]. . . . Had the recycling legislation gone in and met a lot of opposition from union people or [if it had] no support from the unions, it might not have gone anywhere. (personal communication, Oct. 10, 1996)

In this case unionists in the can-making and brewing industries who feared that jobs would be lost if the proposed deposit legislation were enacted averted that threat by agreeing to throw their support behind the proposed recycling legislation, which they viewed as less threatening. Environmentalists sacrificed the deposit legislation that they believed would better protect the environment in exchange for labor support for a recycling program that would partially address the solid-waste problem in the state.

In some instances organizations may even take up causes that were not part of their original missions, in what I refer to as "enlightened cooperation." Enlightened cooperation can occur as a result of organizational learning derived from experience or through ongoing cooperation with other organizations that leads to an internalization of a coalition partner's concerns. To use another example from Wisconsin, unionists in that state, through their close work with environmentalists, developed a greater sense of concern for safety and health in the community, beyond their traditional workplace concerns.

Unions in Wisconsin had worked with environmentalists in support of worker right-to-know legislation, which afforded those in the workplace greater access to information regarding the health hazards associated with the substances they were handling. This fell within the traditional workplace interests of unions. Union involvement in community issues was, however, relatively rare. But after establishing close ties with environmentalists and learning more about environmental concerns, Wisconsin unions expanded their traditional workplace focus and offered their support for community right-to-know legislation. This legislation provided community members with a similar kind of access to information regarding the potential health consequences of the materials being used in manufacturing facilities and other workplaces in their communities.

Promoting expanded access to information regarding possible contamination originating from the firms that employed them posed a potential risk to workers. It is conceivable that concerned community residents, once informed about the toxics being used in a nearby facility, could mobilize and demand its closure, placing jobs at risk. Yet by the time the community right-to-know legislation was introduced, unions had built close working relationships with environmentalists through the Wisconsin Labor-Environmental Network. Their newly developed environmental consciousness and their heightened level of concern for the environment outside the workplace inspired them to move beyond their more limited workplace focus and support this legislation along with their coalition partners. The process of organizational learning will be discussed in more detail in a later chapter. Here we are concerned specifically with organizational barriers to the adoption of new issues that can result from coalition activity.

Organizational considerations can place limits on the kind of cooperative activity that groups are able to undertake. The need for organizational maintenance can clash with the demands of coalition work, depending on the type of cooperation that is being undertaken. The coalition contradiction is completely avoided in cases of instrumental cooperation. Organizations who engage in such cooperation receive the benefits of coalition power without having to depart from their goals in any way. Some research has shown that coalitions operate most effectively when they focus exclusively on issues of overlapping concern (Altemose and McCarty 2001; Brecher and Costello 1990; Cullen 2001; Kleidman and Rochon 1997).

Although instrumental cooperation is, in most cases, not problematic, coalition strategies that involve the need to expand or modify the agendas of the coalition's organizational partners can create problems within the organizations. In the example of compromise cooperation described above, involving recycling legislation in Wisconsin, environmentalists were willing to modify their goals regarding reduction of solid waste through a bottle deposit program in exchange for crucial labor support for a recycling program. This came only after years of failure at getting deposit legislation passed, and in the assessment of environmental leaders, a partial victory was preferable to continued government inaction on the solid-waste issue. Concessions to labor were part of the compromise necessary to gain that partial victory. Environmentalists always maintained that the deposit legislation represented the most environmentally sound policy. But given that labor sometimes serves as the pivotal constituency that determines whether a particular environmental measure will garner the critical amount of support needed for passage, environmentalists in some cases alter their position to lessen the impact on jobs or agree to the inclusion of worker safeguards in an environmental measure.

In Wisconsin, the environmentalists eventually compromised on the bottle bill, but in other cases, environmentalists feel compelled to stand firm regardless of whether doing so means a failure to gain passage of desired legislation because of a lack of labor support. The case cited at the opening of the chapter reflects this sentiment. The NRDC refused to concede to the United Mine Workers on various clean-air measures because, in the words of the NRDC's David Hawkins, environmental "organizations aren't set up to put the union interests first, they are set up to put environmental interests first." The question then becomes what it means for an organization to be "set up" to act in a certain way. Several environmental advocates cite this type of limitation when explaining their inability to adjust their goals in ways that would allow for broader coalition support.

Resistance to the compromises associated with coalition work may come from organizational leaders' personal reluctance to diverge from the cause to which they have dedicated themselves. They may be opposed to expanding or modifying their organization's agenda beyond the core concerns that led to their involvement with the group. Unless some instrumental gain is to be had related to the organization's core mission, these highly committed leaders may be reluctant to stray from the organization's original goals. They may also have concerns about the potential response of the membership to compromises made in the name of cooperation with other organizations. This is how the coalition contradiction manifests itself. Where membership is voluntary, there is always the risk that if an organization diverges from the goals that attracted its members to it, the organization may begin to lose this base of support, thus threatening the viability of the organization itself. This can create pressure to remain narrowly focused on the core goals of the organization. According to Jeffrey Berry in *The Interest Group Society* (1984), "Harmony continues in most groups because of the congruence between members' opinions and the avowed purpose of the organization. . . . [S]taff members are careful not to move the organization toward policy positions that are likely to alienate a significant portion of the membership" (97). Hawkins explains the limitations of the NRDC in general terms, citing the fact that the organization is set up "to put environmental interests first." Some may interpret this position in terms of organizational inertia. Once practices are established, organizations are resistant to change them (Dery 1998; March and Simon 1958; Stinchcombe 1965). This tendency toward resistance to change would certainly apply to social movement organizations and their organizational range, the breadth of goals defined by the organization.

Once their policies are set, groups are unlikely to dramatically change course, simply because their goals and the practices used to achieve them are already established. But leaders of other organizations more directly link their reluctance to diverge from organizational goals to a commitment to their members. Curtis Fisher of the New Jersey Public Interest Research Group describes his organization's approach to environmental issues in this way:

We make decisions based on the environment, and that's my job. When I think about an issue, whether it's nuclear power or dirty coal plants, I have to think about what's in the best interest of the environment. . . . If X chemical is dangerous and the solution is banning it, then the solution is banning it. . . . Our standpoint is from the perspective of a citizen advocate. I know who I represent, these people who give me twenty to thirty dollars. . . . Ultimately, if it's an issue of "do we protect the spotted owl and those old-growth forests or the timber industry?" then yes, we protect the spotted owl. That's what all these people give me money to do . . . that's my job. (personal communication, May 24, 1999)

Fisher describes a general sense of responsibility to the membership of the organization. It is unclear whether a direct concern with the loss of member support drives Fisher's focused commitment to the cause or whether a more abstract sense of responsibility leads him to prioritize core issues even at the expense of labor support for a compromise measure. In either case, the implication is that members will stop paying their dues if he stops doing that which they pay him to do. Thus cooperation with other groups threatens organizational-maintenance needs.

Some environmental leaders are more explicit in terms of the issues they can work on and the alliances they can form based on the risk of alienating the membership of the organizations they lead. In some cases the need to avoid alienating their organization's members imposes very strict constraints that preclude almost any coalition activity with outside organizations, except where interests coalesce perfectly. The issue focus of New Jersey's Clean Ocean Action precludes it from pursuing an agenda involving any issue that does not directly affect the body of water identified by the organization as its mission focus. According to Executive Director Cindy Zipf,

We have very clear guidelines. The goal of the organization is to clean up and protect the waters of the New York–New Jersey Bite, which are the waters that stretch from Montauk down to Cape May. That's our mission, and so anything that is pollution-related or contributing to the degradation of the water quality is something we would immediately act upon. If it's mostly habitat destruction with not much [water] pollution, we would bring it to the board. For example, beach replenishment—dumping sand on beaches—that's traditionally been an exclusive non-pollution-related issue. But several board members are saying that it's not just a destruction-of-habitat issue, that there are contaminants in the sand that are being pumped on the beaches and resuspended into the water column. So we've been just asking questions about the quality of the sand, but we don't get into that issue very much at all. It's a very fine but important line that keeps the organization together. If we get into the beach replenishment issue in a big way, other than to make sure that it's clean, the Boards of Realtors that are members of our group might say, "We can't support that position." (personal communication, May 24, 1999)

Thus, for Clean Ocean Action, habitat destruction on land adjacent to the waters that are the organization's focus is not an issue on which the organization can work unless the contaminants actually get into the water itself. In this case, Zipf cites the limitations imposed by some members of the board of the organization who might withdraw their support if the organization tries to expand its agenda as described.

Having this kind of limitation on its sphere of activity does not necessarily mean that it is impossible for a particular organization to engage in coalition work. In fact, Clean Ocean Action is itself a type of coalition, with a wide range of organizational affiliates. But the cooperation among these affiliates is almost purely instrumental, and straying beyond the specifically defined focus of the group to accommodate other concerns is nearly impossible. In the case of Clean Ocean Action, pollution in the body of water specified as the organization's focus is of mutual concern to all participants, including those affiliated with the real-estate industry. Unlike some other members of the group, realtors may not be motivated by some abstract concern for the environment. Rather, in all likelihood, they share an interest in clean water because of its relationship to property values in the area. Access to attractive beaches will also have an effect on property values, thus for the organization to question beach replenishment on the grounds that it destroys natural habitat would pose a potential threat to realtors' interests and to their continued participation in the group.[6]

These cases indicate that organizational stability and even survival are based in part on retaining a focus on the original central concerns of the organization. This is especially true of voluntary organizations oriented toward the achievement of collective goods, because of the set of incentives associated with the recruitment of members for such organizations. Although the participation of realtors in Clean Ocean Action implies motivation by a material concern, in most cases participation in environmental organizations is based on purposive incentives. Aside from some minor benefits such as newsletters or access to meetings, the motivation to join or remain a member of such organizations is primarily to act in some way that is consistent with one's values and beliefs. To put it in the terms used by New Social Movement theorists, participation in the group provides and reinforces a sense of identity. If I am an environmentalist, I should belong to an environmental organization. If there are grounds to question that organization's commitment to environmental protection, then I may question my involvement in the organization.

The leaders and staff members of environmental groups, in addition to their own concentrated interest in the central mission of their organizations, recognize that it is necessary to remain narrowly focused on the issues that fall within the confines of the stated purpose of the organization for which they work. Compromising on those issues, or even appearing to compromise through working with divergent organizations, can weaken such organizations' appeal to their members and thus threaten their existence.

The effect of straying too far from stated organizational goals can be seen in relation to the dispute over forestry practices in the state of Maine. As described in chapter 4, in the mid-1990s, the more radical wing of the environmental movement in the state initiated a referendum that would have banned forest clear-cutting. The more moderate environmental groups in the state, such as the Natural Resources Council of Maine, entered into a coalition with unions and the paper industry promoting a compromise measure that allowed some forest clear-cutting to continue in order to protect jobs and the economy. The dynamic that emerged between the radical environmentalists, the NRCM, labor, and the paper industry indicates the perils that environmental organizations face as they attempt to protect their organizations while effectively advancing the cause for which they were created.

In this dispute the NRCM was seen by many as having sold out the cause of environmental protection. Critics charged that the organization, in supporting the compromise measure, had been coopted by paper industry sympathizers and that the organization was no longer fully committed to environmental protection. According to NRCM staff member Evelyn deFrees, "For the first time in my tenure on the Council, during the [clear-cutting dispute] there were people actually on the radio from the far left scrutinizing our board members" (personal communication, May 20, 1999). This assault on the NRCM's image as committed defenders of the environment resulted in a significant loss of membership for the organization. People joined the NRCM out of their concern for the environment; joining the organization was an expression of this concern. But when the group appeared willing to compromise on environmental protection to address other concerns, it diminished the appeal of membership in the group. The NRCM board continues to struggle with the issue of banning clear-cutting, as well as other issues that require consideration of the balance between their role as an environmental organization and the need to address the concerns of their allies. According to deFrees,

This is a debate that rages in our organization every time anyone proposes anything new. There are usually two voices simultaneously being spoken that we have to work out. One, we can't be narrow minded because we do our work in a community that is human activity based, and after all, people who don't have jobs can't think of anything else. . . . And there's another voice, equally fair, but from a different perspective, which says, all this is true, but we are an environmental organization. Our mission is to work on environmental issues. There are other groups doing other things that can worry about the other bits and pieces. . . . It's not our job to decide not to do something because it would have a negative effect even though it would bring environmental benefit. This is a policy debate that we go through each time. (personal communication, May 20, 1999)

The NRCM's experience during the referendum on banning clear-cutting demonstrates the organizational peril of veering away from narrow environmental concerns in an effort to address the needs of an organization's nonenvironmental coalition allies. This is where the distinction between public and private organizational goals becomes important. Purposive organizations have little on which to rely in terms of winning and maintaining members other than their claims to effectively promote the cause they have been established to champion (Zald and

Ash 1966). Membership in such organizations is purely voluntary, and members can choose to withhold their support from the organization at any time. Those charged with maintaining such an organization are aware of the potential for alienating members and of their general obligation to advance the specific goals of the organization. In cases like that of the NRCM, the failure to retain such a singular focus created a context in which other organizations siphoned off support by being more narrowly focused on environmental protection.

In that same dispute over clear-cutting in Maine's forests, the flip side of the coalition contradiction plagued the more radical environmentalists. They drew support to their organizations by appearing as the true champions of environmental protection. The referendum on clear-cutting that they were promoting failed, however, because they could not amass the support necessary for its passage. Their narrow focus inhibited their ability to attract the coalition partners needed to advance their goal. The broader countercoalition defeated the measure advanced by the radicals; thus no gains were made in curbing detrimental forestry practices as a result. Although there is no measure of the sentiments of the members of the radical environmental organizations in this case, the failure of these activists to achieve any change in clear-cutting policy probably led at least some of their supporters to reconsider the need for compromise. It may be possible for some fringe organizations to maintain a following based on their ideological purity, but it is unlikely that many organizations can survive over time if they are unable to demonstrate any real progress toward achieving their stated goals.

Environmental leaders and all leaders of voluntary SMOs that are oriented toward achieving public goods face certain constraints in terms of coalition formation, because their public-goods goal orientation and the purposive incentives that they use to attract and maintain members place limits on their ability to alter their agendas. Coalition building will be an option for them only if the goals of other organizations overlap with theirs in a way that allows for instrumental cooperation or if potential coalition partners demand of them only a minimal degree of compromise.

There is great variability, however, in the extent of the scope of the mission that environmentalists define for their organizations. All of the environmental organizations considered in the work presented in this book identify the protection of the natural environment as their primary goal, but beyond that basic commonality, their organizational range varies considerably. Organizations such as Clean Ocean Action have a very narrowly defined set of environmental issues on which they will work. Some, such as the Audubon Society, intentionally limit their focus to particular environmental issues for strategic reasons. According to a spokesperson for the National Audubon Society,

As broadly as we view conservation, we do not have positions on guns, hunting, animal rights, world trade, immigration and any number of other issues that people feel are important and on which other environmental groups hold positions. This is a conscious decision. Our reasons for this are simple: Anything that takes our resources away from our main objectives becomes detrimental to achieving our goals. We're a conservation, education and advocacy group. That's ultimately what we'll stick to in terms of activity and resource allocation.

The limited organizational range of such organizations can stand in the way of coalition building. Potential allies in environmental struggles will often need accommodation or reciprocal support for their own efforts. They will generally not join in coalition with organizations that are so narrowly focused as to preclude any consideration for concerns other than those on their own limited agendas.

Other groups, however, have broader goals or are more capable of expanding their agendas to incorporate the concerns of others, thus allowing them to make greater use of coalition strategies. The Sierra Club, for example, has a platform that includes issues as diverse as electric-utility rate structures, redlining, and the rights of government employees. All of these relate in some way to its environmental agenda; the incorporation of these other issues into that agenda, however, suggests a greater degree of openness in terms of the issues it will work on and, in turn, the coalitions it can join. This is reflected in the greater amount of coalition activity in which the Sierra Club is involved relative to more narrowly focused groups such as the Audubon Society.

Scholars have used various schemes to categorize environmental organizations (Brulle 1996; Gottlieb 1993). For example, Robert Brulle analyzes environmental organizations based on the discourses they use when defining the nature of environmental problems. He identifies six different types of environmental organizations, from preservationist organizations to ecofeminist groups. The key issues of concern in this discussion, however, are the political issues on which a particular organization works. Although there is a great deal of diversity among environmental organizations in terms of their range, one way of distinguishing the relevant differences among them is in the extent to which their concern about the natural environment is tied in with the social ramifications of environmental destruction. This is an important element of an environmental organization's range. Some environmental organizations focus almost exclusively on nature itself, whereas others also include a broader range of environmentally related social issues.

Using this difference in focus as a basis for distinguishing organizations, it is possible to conceive of environmental organizations as lying on a continuum, with distinct ideal types on the extremes of the continuum and other organizations positioned somewhere in between (see figure 5.1). At one end of the continuum are organizations that maintain a very narrow focus on the protection of wilderness specifically. Organizations that reflect this tendency are often referred to as "land use" organizations, although organizations not at all focused on land per se, such as Clean Ocean Action, would fit near this end of the continuum because of their exclusive focus on the natural environment.

On the other end of the environmental continuum would be what are commonly referred to as "environmental-justice organizations." The ideal type in this case would be an organization whose mission focuses exclusively on the way in which environmental hazards affect human health and well-being. The presence of lead water pipes in low-income residential buildings would be an example of an environmental-justice issue, as would dioxin releases from an incinerator located near a residential district. The work of environmental-justice organizations often includes work on toxic substances, and toxics organizations, to the extent they can be distinguished, overlap closely with environmental-justice groups. Environmental-justice organizations are commonly composed of working-class people and people of color who are suffering disproportionately under the effects of environmental degradation.

In the ideal case envisioned for the purposes of demonstrating the range of organizational types, an environmental-justice organization

Land Use --Environmental Justice
Focus on wilderness and   Focus on protecting human health
species protection   from environmental contamination

Figure 5.1

need not reflect concern for the natural environment, in itself, at all. Rather, for such an organization and its members, "environmental contamination provides a proxy for the evaluation, expression and mobilization of human welfare and social justice concerns" (Dreiling 1998, 53). This represents the other pole of the environmental movement, an organization that is essentially concerned with human well-being but focuses on well-being as it is affected by environmental hazards. Of course, in reality no organizations actually fit these ideal types as they are described here. In their rhetoric at least, organizations that lean toward one side or the other typically refer to the social benefits of environmental protection as well as the value of nature in itself. Using these ideal types as a basis, however, we can draw some basic distinctions among organizations at various points on the continuum to see how their organizational range relates to their coalition practices.

The case studies revealed that the organizations that had a more broadly defined agenda or that included the social impacts of environmental destruction in that agenda were more likely to be involved in coalition activity with labor unions. This is not to suggest that the more narrowly focused organizations did not work with others at all. On the contrary, almost every organization had cooperated with others in at least some instances. Virtually any organizations will seek coalition partners if other organized constituents are seen as potential beneficiaries of the same policy that the organization is attempting to advance. For many of the case study organizations, however, coalition activity was limited to instrumental cooperation. This means that the organizations with which they were cooperating were not driven by the same goal, but rather that the goals of both could be achieved through implementation of the same policy and the two groups, in a sense, "used" one another to achieve their ends. John Stouffer of the New York Sierra Club describes a case in which environmentalists agreed to support public enforcement of a union contract in exchange for labor support of land preservation in the area of the contract work: "We had some discussions and it was essentially a quid pro quo where they write us a letter saying they support preservation of the land, and we write them a letter saying we support the labor thing. In a situation like that, once either party gets what they want, they are gone. So it's not a situation where there is a high value in cooperation" (personal communication, May 12, 1999).

As can be seen, there need not be any concern at all with the ultimate goals of the other organization in a coalition effort beyond securing its aid in advancing a policy that happens to benefit both organizations. Even very narrowly focused organizations are capable of engaging in this kind of coalition activity. In New York, for example, the Audubon Society director had coordinated activities with building-trades union leaders around the cleanup of Long Island Sound. The Clean Water Jobs Coalition that they formed focused on attaining federal funds to upgrade existing sewage treatment plants that discharged materials into the sound and to construct new ones. This would obviously create many jobs in the building trades, fulfilling the unions' goal of creating jobs and advancing the interests of their members, while at the same time achieving the Audubon goal of reducing water pollution in the sound.

But beyond cases in which the interests of two organizations instrumentally align, organizations that have a very limited range are less capable of coalition activity than those with broader ranges. In some cases the restricted focus of such organizations can turn potential coalition partners away even when the two have mutual goals. Leaders of the New Jersey Audubon chapter (which is technically independent from the national Audubon organization but has a similar land use focus) attempted to enlist urban leaders in the organization's campaign to reduce sprawl and preserve natural areas peripheral to urban areas. According to Director of Conservation Bill Neil,

We had outreach to county organizations and urban leaders without a lot of success from the urban side. [We saw them] as natural constituents in any type of attempt to do land use. We thought urban areas are not getting the jobs. We are still losing population in many cities. We are logical allies. You need allies to do this work, but this has been a pretty miserable failure. It's very clear the urban leaders do not see that as their priority. (personal communication, May 23, 1999)

Given the limited scope of activity among groups like the Audubon Society, coalition mobilization is likely for such groups only when interests among the groups are perfectly aligned on a particular issue. In some sense this specific case may represent a failure on the part of urban organization leaders to recognize the potential benefits of working with the Audubon Society. But Audubon's exclusive focus on land conservation did not inspire support from communities that see social issues as their central concern. Land use organizations are limited in their ability to engage in coalition activity beyond the purely instrumental type. The more narrow their focus, the less coalition potential they will have. The incorporation of additional concerns into their agendas that is often associated with other coalition activity would push them beyond where their limited agenda permits them to go.

In contrast, organizations that have provisions that more directly concern the social ramifications of environmental degradation are more capable of engaging in coalition activity. Those that I refer to here as "broad environmental groups," whose range encompasses both land use and socially related environmental issues, are less constrained in the issues their agendas can accommodate, allowing such groups to engage in more coalition activity than those with a narrower range. The breadth of their platforms, in a sense, provides their leaders with the cover they need to justify the incorporation of more peripheral issues, while enabling them to continue fulfilling their obligations to their members. Since these issues are to some extent already part of the organizational platform, decision makers in these organizations are less concerned about member withdrawal as a result of coalition activity. And of course, actions taken in respect to these issues are especially defensible because coalition support often makes real gains on the organization's core issues more likely.

Environmental-justice groups have an organizational range that makes them more likely than other environmental organizations to enter into cooperative relations with organized labor (Dreiling 1998) (see table 5.1). Because their range includes both a social and an environmental component, these organizations are more capable of coalition formation with labor unions than are land use organizations. Similarly, environmental-justice organizations, because of the environmental component of their agendas, may be more capable of working with land use organizations than are labor unions. On the whole, however, most

Table 5.1
Environmental SMOs: Organizational range and coalition activity

Organization	State	Type	Coalition activity
Forest Ecology Network	ME	Land use	Low: Some labor outreach, but radical and resistant to compromise, no participation in formal coalition
Maine Greens	ME	Broad environmental	Low: Some labor outreach, but radical and resistant to compromise, no participation in formal coalition
Natural Resources Council of Maine	ME	Broad environmental	High: Close ties with labor, participate in coalitions
Sierra Club	ME	Broad environmental	Moderate: Some unilateral cooperation with unions, little coalition involvement
Audubon	NJ	Land use	Low: Some unsuccessful outreach to labor for instrumental purposes
Clean Ocean Action	NJ	Land use	Moderate: Regular contact with a small number of unions
New Jersey Environmental Federation	NJ	Broad environmental	High: Very close ties with industrial unions
Sierra Club	NJ	Broad environmental	Moderate: Some regular contact with certain unions
Audubon	NY	Land use	Low: Some instrumental cooperation with the building trades, minimal state coalition involvement
Citizens Environmental Coalition	NY	Toxics	High: Central environmental group in the labor-environmental network
Environmental Advocates	NY	Broad environmental	Moderate: Some participation in labor-environmental network and work on some other issues with unions
Rainbow Alliance for a Clean Environment	NY	Toxics	High: Much participation with the labor-environmental network and close ties with local union

Table 5.1
(continued)

Organization	State	Type	Coalition activity
Sierra Club	NY	Broad environmental	Moderate: Some participation in labor-environmental network and work on some other issues with unions
Audubon	WA	Land use	Low: Some informal ties with labor, no regular contact
Sierra Club	WA	Broad environmental	Low: Some instrumental cooperation with tugboat union, some informal ties with others, no regular contact
Citizens for a Better Environment	WI	Toxics	High: Regular participation in the labor-environmental network
Sierra Club	WI	Broad environmental	High: Regular participation in the labor-environmental network

Note: Organizations were characterized as being one of three types: land use, broad environmental, or toxics. Land use organizations are those that place most of their emphasis on wilderness and species protection. Broad environmental groups place emphasis on a range of issues including wilderness protection and toxic substances. Toxics organizations place most of their emphasis on industrial pollution and the health hazards it poses. Although there is considerable overlap within these categories, it is possible to differentiate among organizations in each of them. For example, toxics organizations rarely devote significant effort to issues such as public-wilderness purchases. Similarly, land use organizations do not often concentrate on hazardous-waste issues, unless they threaten natural areas. Broad environmental organizations have an issue range such that they are active on both kinds of issues. Progressive citizens' organizations whose work includes environmental issues are not included in this table, since they are not properly considered "environmental" organizations. Their role, however, will be considered within the discussion. Three sources of information were used to place groups into categories: interview data, organizational literature, and secondary sources. Levels of coalition activity were determined based primarily on interview data, with some supplemental information from organizational documents. Environmental groups that participate in coalitions that include unions or that maintain unilateral cooperation with several unions were designated "moderate" or "high" in terms of coalition activity depending on the extent of those ties. Those with no labor contact or for whom there is limited contact for purely instrumental purposes were categorized as "low" in terms of coalition activity.

land use organizations are fairly constrained in their ability to act in regard to issues that deal exclusively with social justice. Broad environmental organizations have a range that encompasses all types of environmental issues, including some that lean toward the social-justice end of the spectrum. They are more capable of coalition work with unions than narrow land use groups, but less so than explicitly environmental justice or toxics groups.

Organized Labor

Later we will examine the issue of a voluntary organization's ability to *expand* its agenda and make the accommodations that will allow it to engage in more coalition activity, but first let us consider the limitations faced by organizations of the other movement of interest in this book, labor. In contrasting the organizational range and structural limitations imposed on labor unions with those affecting environmental SMOs, it is important to distinguish the relatively centralized structure of the labor movement from that of the highly differentiated and diffuse environmental movement. The labor movement in the United States has a dual three-tiered structure, with one hierarchy consisting of "international" unions composed of workers associated with a given industry[7] and a second composed of union federations under the leadership of the AFL-CIO (see table 5.2). Some workplace unions are completely local and independent of any international body, but most individual workers in unionized fields and industries will belong to a union local that is affiliated with an international union. Union locals are typically based in one industry or in one company. There may be other regional levels within the international (commonly called "districts") that are situated between the locals and the central international union; these regional levels make up the third of three tiers of union organization.

The second organizational pillar of the labor movement is the AFL-CIO, a national federation that also has affiliated local and state organizations. Central labor councils (CLCs) encompassing a city or local region are the lowest level of organization of the AFL-CIO. Union locals within that city or region (which, again, are affiliated with an international union through the other hierarchy) compose the membership of the CLCs. The next level is composed of state federations that include

Table 5.2
Organizational structure of the labor movement in the United States

	Individual unions	Union federations
National level	*International unions* National unions traditionally organized around a particular industry such as the United Auto Workers and the United Mine Workers	*AFL-CIO* The national union federation with which international unions are affiliated
State/regional level	*Districts* State or regional level body of the international with which union locals are affiliated	*State federations* State-based federations with which local unions from different international are affiliated
Local level	*Union locals* The local organization representing workers at their workplace	*Central labor councils* The local federation with which local unions from different internationals are affiliated

Note: Based on Bobo, Kendall, and Max 2001, 207.

union locals and CLCs from within the same state. The state federations and the international unions are affiliated with the national AFL-CIO, the highest level of the encompassing organization.

Particular procedures may vary from union to union, but in most cases local union leaders are elected by the local membership. State federations and CLCs are the labor focus in the study presented in this book, although some representatives of union locals and internationals were also interviewed in the course of the research. The state federations are key players in coalitions with environmentalists because they are primarily responsible for promoting the political interests of union workers at the state and local levels. State federation and CLC officeholders are elected by the affiliated locals based upon a formula that takes into account the size of the membership in each particular local that belongs to the CLC or state federation.

Because CLC and state federation leaders are elected by the representatives of local unions, they are accountable to the membership of the locals. In addition, although while some internationals require that their locals affiliate with central labor bodies, for many union locals, affiliation with the CLC and state federation in their area is voluntary. This again serves as a force making the representatives at the CLC and state levels accountable to the locals. This creates an effect similar to that which environmental leaders experience because of the voluntary nature of membership in environmental organizations. That is, union representatives at the CLC and state federation levels are accountable to the members of the union locals through some form of democratic process and also through the threat of exit. This can inhibit the ability of the leaders in both types of organizations to pursue strategies that depart in any way from the interests of the members, including those that may be called for by coalition affiliation.

There are at least two key differences, however, between the position of union leaders and the situation faced by environmental advocates. The first is that workers and local union leaders, unlike members of environmental SMOs, are limited in the options they have for representation. An individual concerned about the natural environment can choose to affiliate with any number of environmental organizations to advance the goal of protecting that environment. Workers and union locals do not have this option. Because of the more centralized nature of the labor movement (compared to the much more diffuse environmental movement), local unions have only one CLC and one state federation with which they may affiliate. They can either have representation at these levels or not, but if they choose representation, they have no choice as to who represents them. It would be possible for a local to withdraw from these central bodies and attempt to promote its own political interests directly, but only the largest of locals realistically have the option of doing so, and even then, those that choose to do so may lose a great deal in terms of the broader support that a CLC or a state federation can offer.

This situation parallels that of an individual union member and their local union. Union membership is not required; an individual surrenders their voice in representation, however, if they choose not to join the union.[8] The threat of withdrawal from the union is not significant if no viable alternative to union membership exists; thus individuals have little incentive to quit their unions even when the unions adopt positions contrary to the individual's views. It is possible for workers collectively to disaffiliate with their union; this is a major undertaking, however, that is likely to be entered into only under extreme circumstances. This organizational structure provides union leaders with greater flexibility in terms of the issues on which they can decide to work. Compared to environmental leaders, who have to be constantly concerned about the appearance of diverging from the core focus of the organizations they represent, union leaders face a relatively weaker constraint in this regard.

The second major feature of the position of labor union leaders that differs from that of leaders of environmental organizations is the range of issues on which they are empowered to work. Individuals join environmental organizations specifically to support the protection of the natural environment. Unions, in contrast, are charged with promoting the interests of workers across any number of issues. This is especially true of labor federations. Local unions are responsible for defending the workplace interests of their members, and at this level union members are likely to expect direct private return in exchange for their membership (this issue is dealt with more extensively in chapter 8). Labor federations, however, carry out the political work of the labor movement generally, and at this level we often see the pursuit of goals that have only a tangential connection to the workplace interests of union members. When Ed Donnelly, the director of legislation at the New York State AFL-CIO, was asked whether the federation took positions on issues not directly related to workers, such as education or welfare or the environment, his response was, "Those issues are related to workers" (personal communication, May 10, 1999). This can be seen as a sharp contrast to the position of environmental leaders, who are limited in the connections they can draw between other issues and environmental protection.

Despite the perception of unions as having a narrow focus on the material interests of their members, union leaders at higher levels of the organization often have a free hand to address a wide range of issues that may bear no direct relationship to the workplace. Some of the union leaders interviewed in this study explained that their job was to advance the interests, broadly construed, of workers—union members and non-members alike—and their families. To the extent that environmental protection is seen to be in the interests of workers, union leaders are free to pursue that goal. Rank-and-file members of the union are unlikely to oppose such action provided jobs or other core interests are not threatened by it.

Yet despite the broader range of issues that can fall within the purview of union advocates, traditionally, in the United States, many unions have limited their scope of activity to issues concerning the wages, hours, and working conditions of their members.[9] This tendency relates to the discussion in chapter 3 of "business unionism" as compared to "social unionism." Unions' exclusive focus on securing benefits specifically for members is characteristic of business unionism, an approach to worker organization in which the union operates essentially as a business that provides services to members in exchange for dues. The business union approach was embraced by the skilled craft unions that made up the early American Federation of Labor. Their primary goal was to improve wages for members by restricting access to work skills required to engage in the crafts they represented. Business unionism today retains this exclusive focus on workplace issues and generally shies away from political activity, especially that which diverges from the primary areas of member servicing: wages, hours, and working conditions.

The narrow organizational range characteristic of business unionism contrasts with the social unionism advanced by many of the industrial unions of the former Congress of Industrial Organizations. Social unionism is generally characterized by efforts to advance the interests of workers as a class, as opposed to limiting advocacy to that on behalf of union members alone. It also involves a broader organizational range in terms of the goals the organization pursues, which extend beyond issues involving wages, hours, and working conditions that pertain specifically to the workplace (Seidman 1994). The social-unionism approach is also focused to a greater degree on mobilizing workers and expanding union membership, as opposed simply to servicing existing members. Most of the U.S. labor movement adopted the business union strategy when the AFL and CIO merged in 1955, and the CIO union radicals, who were more inclined toward the class-based approach, were purged from the union's membership with the advent of the Cold War.

The conservative business union approach served the interests of many union members well for several decades. Under the social contract of the postwar years, union workers were cut in on the nation's prosperity in exchange for labor peace. Although it benefited many, this strategy also led to stagnation within the labor movement. Unions, prioritizing the interests of existing members, dedicated few resources to recruiting or organizing new members, leaving out large segments of the workforce and generally weakening their strategic position. Changes within the economy, in particular the decline in manufacturing employment, coupled with the recession of the 1970s and the anti-union assault by corporate and government leaders in the 1980s, led to continuously declining union density.

Although most unions in the United States embraced the business union approach as of the middle of the last century, some of the industrial unions, such as the UAW and the OCAW, maintained elements of a social-union orientation. In many cases their political work incorporated concerns well beyond those of their particular members. The broader organizational range of these unions created more opportunities for them to work in coalition with other types of movement organizations, including environmental groups. This can be contrasted with the approach of the building-trades unions, which more than other union sectors have maintained a business union orientation as laid out by the early AFL. They have tended to focus narrowly on the work-related interests of their members, and this limited organizational range has consequently limited their coalition activity (Dreiling and Robertson 1998; Siegmann 1985).

To some extent union federations will, by definition, be involved in a broader range of issues than their individual affiliates. The specifics of wages, hours, and working conditions are typically settled through collective bargaining, which is carried out by union locals or the international. The job of CLCs and state federations is to address the political interests of workers that extend beyond those that can be included in a collective bargaining agreement. Contrasting social and business union strategies can also be seen at this level, however, and the findings of this study confirm what earlier research suggests about union willingness to work with the environmental community: Social unionists tend to engage in more coalition work, whereas business unionists refrain from this type of involvement.

Relations between the environmentalists and unions are poorer in New Jersey than in many other states. The building trades also have a significant influence within the state AFL-CIO. A spokesperson for the federation described the political approach of the organization:

At the AFL-CIO our primary message is looking out for employees and workers' rights, and while the environment is part of that, it's a bit more of an attenuated part than, let's say, workers' compensation, which has an obviously much more direct effect on you as an employee. And when you start to branch out to issues where it's not crystal clear why you are interested, you start to dilute your message. . . . You sometimes have to focus on one very narrow thing, or you risk losing that by branching out to other no less important things. But there are other groups who have to carry that part of the burden.

The business union orientation is evident in this description, and the similarity between the approach described and that of some of the land use environmentalists demonstrates that narrowness of organizational range can be found in both sectors. The same labor representative went on to explain how the federation makes decisions regarding what position to take when an issue arises:

A lot of it is self-explanatory. For instance, let's say there is a construction bill to pump $25 million into construction of a new facility. That means construction jobs, and our building trades are active locally. We have a lot of building trades unions, whether it's electricians, carpenters, plumbers, or sheet metal workers. And obviously what's good for the construction industry is going to be good for union members. Those issues are like softballs. You know which side you have to be on. But sometimes you come to issues where it's not so cut and dry. Does it affect enough of our members or even any of our members? For example, [an issue] might only affect private bus drivers as opposed to unionized bus drivers.

The focus on the servicing of members as it relates to jobs and the workplace is obvious from these comments. Even job-related issues that do not directly affect members may not be considered within the purview of the federation, according to the rationale presented. Contrast this statement with that of Ron Judd, executive secretary-treasurer of the

King County Labor Council in Washington, as he explains the role of his organization:

The mission is to create and organize for a progressive social economic agenda that promotes working families. That's really what we're about. We go much farther than just hours, wages and working conditions. I like to think of us as dealing with the social and economic conditions of all workers within our community. And that can be of health care or welfare reform or the siting of an incineration plant or whatever that might be. . . . In fact I think that one of the problems that the labor community has had in this country is that they really haven't articulated what the values of having a strong labor movement is socially or economically or for the community. We have been too narrowly focused in the past on collective-bargaining agreements and we have not been focused enough on the community. (personal communication, March, 29, 1999)

The contrast in the way in which these labor leaders describe the missions of their organizations can help us understand the failure of labor-environmental coalitions to form in some states, whereas in others labor unions and environmentalists work closely together. New Jersey has the poorest quality of labor-environmental relations of those states examined in this book. Numerous conflicts between unions and environmentalists were identified by leaders on both sides, and aside from a few instances of cooperation with particular industrial-union locals, relations between environmentalists and state AFL-CIO leaders are generally poor. The state AFL-CIO tends to be dominated by the building trades, and they follow in the craft union tradition of business unionism with its correspondingly narrow range. In Washington labor-environmental relations are not optimal, in part because of the legacy of the timber conflicts of the 1990s; however, the state AFL-CIO and the King County Labor Council in particular have established some positive ties with the environmental community, and they have cooperated with environmental organizations on a number of efforts. The broader social-union orientation that these labor federations embrace allows for the incorporation of environmental concerns into their agendas and opens the door for coalition work.

The tendency for greater coalition activity to correspond with the organizational range of the labor federation in the state can be seen in each of the cases examined. As noted, in New Jersey, where the state federation's approach is most reflective of the business union orientation, relations between labor and environmental organizations are poor, and coalition activity with the AFL-CIO does not exist. This built-in union resistance to coalition activity is compounded by the fact that several of the most active environmental organizations in the state, such as the Audubon Society and Clean Ocean Action, have adopted a fairly limited organizational range. Some other environmental groups in the state with broader agendas that incorporate elements of environmental justice do have close working relations with some unions outside of the state federation framework. This is the case, for example, with the New Jersey Environmental Federation and the New Jersey Industrial Union Council, a federation of industrial unions separate from the state AFL-CIO. New York and Wisconsin, where the state federations both reflect more of a social-union approach, have the closest labor-environmental relations of those states examined. In Maine, the state federation has a broad social-union approach, and its relations with some segments of the environmental community are very close. Some of the more radical environmental organizations, however, which are more focused on environmental concerns and argue that compromise is not necessary to address the interests of labor, have clashed with union leaders in some instances.

The cases discussed above demonstrate that those organizations (and leaders) who conceive of the role of organized labor more broadly are in a better position to utilize coalition strategies than those who take a narrower view of labor's role. The goals of coalition partners not only can be accommodated but may be part of the overall mission that union leaders consider to be within their purview. This is not to suggest that labor unions will naturally ally with any movement organization seeking some public good. With any worker organization there will still be a tendency to require that there be some link between the well-being of workers and coalition goals when participation in a coalition is being considered. This is especially true if there are dimensions of the issue around which the coalition is being built that could be seen as in some way threatening to the interests of workers. But again, even in those cases, those worker representatives who conceive of their role in broad terms are likely to weigh the balance of benefits and drawbacks in coalition participation. In Washington, labor unions have actually opposed some building projects on the grounds that they pose an environmental threat to the community, despite the fact that those projects would offer jobs for union members. This is in contrast to New Jersey, where, if a project will provide jobs, "you know which side you have to be on," as the New Jersey AFL-CIO spokesperson put it. Making a trade-off between potential benefits and potential drawbacks of coalition participation will be given more serious consideration by those union leaders who include community concerns in their analysis. If environmental organizations are in a position to adjust their demands to accommodate the interests of workers, coalition activity between labor unions and those organizations is most likely.

Figure 5.2 depicts graphically the way in which the organizational range adopted by different labor and environmental organizations can result in various degrees of overlap in the organizations' goals and agendas, which then either allows for or prevents coalition activity. The organizations from each sector that maintain a more narrow range, land use environmental organizations and business unions, have little overlap with other organizations. Land use groups focus exclusively on wilderness protection and other specifically environmental concerns. Social issues are not part of their agenda. Similarly, business unions focus on a narrow range of primarily material issues that relate specifically to their members. They typically do not address environmental issues or other broad public goods. Because of their narrow range, these organizations engage in instrumental coalition activity only around policies that promise distinct benefits to them. Broad environmental organizations and social unions, on the other hand, have a larger range that incorporates some social and environmental issues. Having areas of mutual concern, as signified by the overlap in range between these groups, allows for more cooperation between them. In addition to opportunities for instrumental cooperation, there is a greater likelihood in such situations that compromises can be reached on a particular issue that satisfy both sides or that cooperative work can be carried out on issues of mutual concern. Because they bridge the divide of social and environmental issues, environmental-justice organizations and those that work on toxics issues are in a position to maintain cooperative relations with those on the labor and environmental sides of the spectrum.

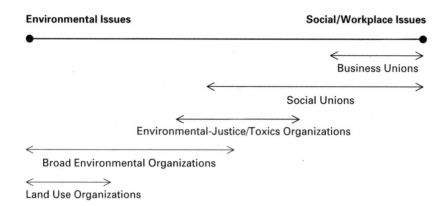

Figure 5.2
Organizational range of labor unions and environmental SMOs
Note: Organizational range refers to the scope of the goals sought by various movement organizations. Business unions and land use environmental organizations have a more limited organizational range, with unions focusing narrowly on work-related issues and land use groups focusing exclusively on wilderness and species protection. Broad environmental organizations are those that seek to protect the environment but also incorporate social issues into their agendas. Social unions not only seek to advance the interests of their members as workers but also adopt goals that apply more broadly to members of the community or the public as a whole. Environmental-justice organizations focus on the social ramifications of environmental issues. Organizations that adopt a broader organizational range or incorporate both social and environmental issues into their agendas are generally more capable of alliance formation. Those that maintain a narrow focus on either environmental or social issues may be capable of instrumental cooperation when policies allow for the satisfaction of diverse concerns, but they are more constrained in their ability to work with others when such efforts would require them to expand beyond the limited range of issues that they take as their normal concern.

Expanding Agendas and the Growth of Labor-Environmental Coalitions

Figure 5.2 categorizes various types of organizations according to the breadth of their orientations. Although the actual practices of a particular union or environmental SMO are unlikely to fit perfectly within any one of the categories depicted in the figure, the tendencies of the various categories of organizations in terms of coalition activity are apparent. Organizations that tend toward a broader conception of their roles are those most likely to engage in coalition activity. Narrowly focused business unions and land use organizations are less likely to be able to accommodate the concerns of coalition partners and are therefore less likely to participate in coalition activity, although instrumental alliances are still conceivable for any type of organization. As demonstrated by the coalition in New York built around the cleanup of the waters of Long Island Sound that included the building trades and the Audubon Society, even land use organizations and business unions may find that cooperation to have certain policies implemented can serve their distinct interests. We can expect more in the way of coalition activity, however, from those organizations with more encompassing orientations.

Recent changes taking place within the labor and environmental movements may actually facilitate further expansion of the ranges of organizations in these movements and thus increased labor-environmental cooperation. Michael Dreiling (1998) argues that pressure from grassroots organizations is forcing others to alter their approach to issues outside their traditional range. In his analysis of labor and environmental cooperation around opposition to NAFTA, he argues that recent intermovement alliances "stem from the way in which these groups responded to, or were redefined by, movements from below, i.e. social movement unionism and environmental justice, respectively" (53). He argues that "decentralized, local, militant, and democratic organizations tend to operate with greater levels of inclusiveness" and that grassroots environmental-justice and social-union mobilizations in opposition to NAFTA forced reform within national labor and environmental organizations.

Yet, it is unclear whether, on the whole, grassroots organizations always tend toward inclusiveness in ways that would allow for inter-movement collaboration. In fact there is some evidence, from the study reported in this book and other studies, to suggest that organizational leaders are better positioned than are rank-and-file participants to adopt a broad vision of the organizations' goals (Rucht 1999; Sabatier and McLaughlin 1990). Union locals often focus narrowly on member job protection and workplace concerns, whereas international and federation leaders are capable of identifying larger issues that affect the membership as a whole, as opposed to particular individual concerns. Similarly, in the environmental movement, some of the most grassroots environmental organizations are also the most narrowly focused on environmental concerns and thus the most lacking in the propensity to make the compromises necessary to accommodate coalition partners. In Maine, for example, the most grassroots based of the environmental SMOs, the Greens, have also been the most militant in refusing to compromise in ways that would accommodate union concerns about job loss.

In addition, in at least some cases, labor-environmental collaboration at the state and local levels has been facilitated from above. The long-standing Wisconsin Labor-Environmental Network was the product of a national meeting during which top labor and environmental leaders adopted a strategy of state coalition building (Obach 1999). But regardless of whether grassroots organizations are themselves more capable of collaboration, the recent developments identified by Dreiling, including mobilization by new sectors within the environmental movement, have created pressure on the national organizations to expand their agendas in ways that make collaboration with other organizations more likely.

The same can be said of developments within the labor movement. As noted previously, after almost fifty years of what is best characterized as a business union orientation within the AFL-CIO, the leadership of new AFL-CIO president John Sweeney has redirected the federation toward a more activist social-union approach. There is currently a greater focus on organizing new members and on rebuilding the labor movement's political strength. This has included outreach to other potentially supportive constituencies, including the environmental movement. In Sweeney's words, grassroots political and legislative work highlight the new emphasis our movement is placing upon relationships with community institutions—local governments and their elected leaders, religious groups, civil and human rights organizations and other allies. . . . [W]e are putting unprecedented numbers of activists on the streets to support organizing and bargaining. . . . [W]e also are developing ways to make our communities better places to live and work. We are helping them hold employers accountable for their treatment of workers and the environment. . . . And because our communities need us in new ways—just as we have new needs for community alliances—we redoubled efforts to mobilize our members and intensified our local outreach, involvement and coalition building.

The extent to which all union sectors will embrace the social-union approach advocated by Sweeney is unclear, as is the commitment of the new AFL-CIO leadership to policies that will truly empower the working class (Tillman and Cummings 1999). Some of the service-sector unions (Sweeney rose through the ranks of the Service Employees International Union, where he was president before being elected to lead the AFL-CIO) have fully embraced the new strategy, and a number of industrial unions have incorporated aspects of social unionism into their missions all along (Johnston 1994). By some accounts the building trades remain recalcitrant in their business union orientation, shunning work with other organizations. But to the extent that the new social unionism represents a growing development within the labor movement, it suggests that labor-environmental coalitions will become more common.

The environmental movement is likewise undergoing a transformation. Waves of declining membership in some of the largest environmental organizations in recent years have led some to reconsider their approach to environmental issues. The rise of the environmental-justice movement and the critique it makes of mainstream environmental groups have also led to a reevaluation of goals and approaches among those organizations (Bullard 1993). In the face of this critique, many environmental organizations are attempting to shed their elitist image by expanding their agendas to give more consideration to the social repercussions of environmental protection. According to Jeff Jones of Environmental Advocates, based in Albany, New York:

We are trying to expand our approach to the pollution issues by, at the same time, working on some of the primary human health issues that are, in one way or another, caused by the pollution we are trying to prevent. . . . New York state is the epicenter of the asthma epidemic. We should be in touch with people who are fighting to come up with a comprehensive approach to the asthma epidemic. That makes a much broader alliance. . . . We also have an analysis that New York state is changing demographically. Minorities are going to be the majority in New York within the next generation. So, to those of us who are living in upstate New York and love the environment, and have done a pretty good job of protecting the environment of upstate New York, if we want to continue doing that, we better have good working relationships with the people of New York City, who generally experience the environment as a hostile force. Instead of the traditional limits that existed in the past, where they are talking about people from the Upper East Side contributing money to . . . protect the Adirondacks, we are talking about building a political force that deals with issues like what you do about all the garbage in New York City and asthma and things that are really impacting people's health. (personal communication, May 7, 1999)

In addition to this growing concern for expanding the base of the environmental movement and incorporating the social impacts of environmental degradation into the movement's agenda, as described in chapter 2, environmentalists have also had to confront arguments regarding the economic impacts of environmental protection. Industry claims of jobs lost (or potentially lost) through the implementation of environmental-protection policies and worker opposition to environmental measures have forced environmentalists to confront arguments about the economic ramifications of environmental regulations. In many instances environmental organizations now conduct economic analyses of the effects of the policies they advocate and develop strategies to ensure that job loss does not result from the implementation of those policies. For example, a pamphlet (n.d.) from an environmental group working on the protection of forest land in Maine ("RESTORE: The North Woods") touts the 20,000 new jobs in "clean service industries" such as "education, trade, information, finance, health care and other well paying professions" that would be generated if the group is successful in having a national park created in northern Maine. It contrasts this potential creation of jobs with the diminishing number of jobs provided by the increasingly automated timber industry. In another example, the Sound Economy program promoted by the People for the Puget Sound in Washington incorporates the creation of habitat restoration jobs for displaced timber workers into its forest protection plans. The group has also conducted analyses on the economic harm to industries that rely upon the sound that results from environmental damage in the region.

High-profile battles such as that over the spotted owl in the Pacific Northwest and conflicts over global climate change served as the impetus for the new strategic positions being staked out by environmental organizations. According to David Hawkins of the NRDC, "the environmental community . . . is realizing that the best way to protect the remaining old-growth forests is to come up with alternative economic visions for the regions in which those forests are located. And not just have a vision of what the ecosystem protection component of the future looks like, but also what the people that live there are going to do" (personal communication, June 4, 1999).

These positions are very different from the arguments made by environmentalists in the 1970s in favor of a zero-growth economy, a position that few environmental organizations voice today. Although the incorporation of economic analyses into environmental protection plans may represent an effort to preempt job loss arguments (and some environmental leaders do acknowledge that economic issues enter into their strategy considerations only *after* environmental concerns have been analyzed), the trend toward explicit consideration of employment concerns in environmental agendas nonetheless demonstrates an expansion of the exclusive wilderness protection focus that has been dominant in earlier periods.

Essentially what we are witnessing is an expansion of agendas on the part of both unions and environmentalists. Rick Bender of the Washington AFL-CIO suggests that there has been something of a change within the environmental movement.

At least now [environmentalists] are taking a look and trying to respond to the economic impacts of some of their measures. But we have also taken a step too towards that . . . a good environmental project also means good jobs for us. So there are ways that both sides have had to come a little more towards the center in regards to how they look at the economic consequences. I mean if we see a job that has some real strong environmental problems, we may not support that job, even though it might put our people to work. (personal communication, March 26, 1999)

The trend toward expanded organizational agendas among both types of organizations would suggest that labor-environmental coalitions will become more common in the future.

Conclusion

The political structure in the United States tends to create a fragmented interest group system in which numerous social movement organizations form to advance a narrow set of goals. Organizations are induced by the political structure to focus on a limited range of concerns, such as the protection of the natural environment, and within that, to target even more specific issues. This tendency became even more prominent when the social and political changes of the 1960s spurred the creation of a multitude of organizations addressing a myriad of specific issues. But although this mobilization may have contributed to opening up the political process to greater citizen participation and challenged the control of elite political insiders, the sheer number of organizations that formed in that decade had the ironic effect of limiting the power of individual groups. Although coalition formation is a standard practice within democratic systems generally, the more densely populated political landscape that emerged at the end of the 1960s made coalition formation among movement organizations an essential strategy for achieving their organizational objectives.

Yet SMOs face certain constraints when attempting to forge coalitions, depending on the types of goals they are seeking to achieve and the range of issues they address. Whether they are striving for public versus private goods interacts with the scope of their missions, allowing some to join more easily with others in coalition activity. Voluntary organizations seeking public goods, like environmental SMOs, must find a balance between maintaining their organization and addressing the issues that serve as the reason for their existence. They risk losing their appeal if they are perceived as lacking focus or if their more encompassing agendas alienate potential supporters. As organizations striving to achieve a public good, the only strong incentive for participation that they can offer to members is the sense that they are contributing to a cause in which they believe. Thus to distinguish themselves and attract supporters in the densely populated political space, they seek to address increasingly specific causes, but at the same time they must embrace an agenda with enough breadth and flexibility that it overlaps sufficiently with the agendas of other organizations to allow for coalition activity and the political efficacy that comes with it; thus arises the coalition contradiction.

Labor unions are less constrained than environmental SMOs by the coalition contradiction because they enjoy a unique status. As the exclusive representatives of the interests of certain workers, they face, to a much lesser degree, the kind of competition that compels a narrow organizational range among environmental groups. This is especially true at the federation level, where political actors are not directly tied to specific workplace concerns and are thus more capable of integrating a broad range of issues into their agendas then union locals. The protection of the natural environment is sometimes among those issues. But federation leaders are likely to face some resistance from their affiliates, as they seek to incorporate a broader range of issues into their advocacy work, especially if those issues may in some way threaten the basic economic interests of some workers. In recent history unions in the United States have operated primarily to benefit their particular members. The business union tradition is still strong in some sectors, and union federations may be forced to abandon environmental allies if actions undertaken with those allies threaten traditional interests. Thus narrow land use environmental organizations and business unions rarely find the opportunity to join with others for mutually beneficial political action.

Yet some organizations in both the labor and environmental sectors have managed to maintain a broader organizational range. Those labor organizations that incorporate into their organizational agendas some amount of environmental concerns and those environmental organizations that embrace social concerns to some degree have an overlap in organizational range that allows for coalition activity. Social unions and broad environmental organizations have interpreted their missions in a way that enables them to identify one another's issues as being within their range of concerns. Environmental-justice organizations and anti-toxics groups also span the divide between limited environmental and social concerns.

And there is some evidence to suggest that this divide is shrinking even between those organizations with a traditionally limited range. Pressure exerted by those groups with a broader focus on their more narrow counterparts is generating a range expansion that will likely allow increased labor-environmental cooperation in the future. Social-union advocates, like the AFL-CIO president, John Sweeney, have actively encouraged unions to adopt a more inclusive strategy and to seek coalition partners such as environmental SMOs. Likewise, traditionally narrow environmental organizations have been pressured by environmental-justice organizations and stung by defeats associated with their neglect of social and economic concerns, resulting in a broadening of the scope of their advocacy work. The actual process of how this range expansion occurs at the organizational level is the subject of the next chapter.

6

"I Had a Lot to Learn about Those Issues": Organizational Learning among Unions and Environmental SMOs

Dominick D'Ambrosio was born the son of a West Virginia coal miner. From growing up around mines and mine workers, he learned a great deal about the struggles of workers and even about the health threats posed by the dangerous work in the mines. But he was never very informed about environmental issues. After graduating from high school, he went to school for broadcasting while working in a factory to pay his bills. After a short stint as a professional radio broadcaster he was offered a staff position at the Allied Industrial Workers International Union, where he would eventually rise to the position of president. Environmental concerns were remote in his mind, and he seldom made the connection between those concerns and the concerns of those he represented during his work as a union leader. But that began to change in 1981 when he was invited to attend a national meeting organized by Howard Samuel, of the AFL-CIO's Industrial Union Department. Samuel had previously met with Sierra Club President Mike McCloskey and other national environmental leaders to strategize about how to defend against the threat of regulatory rollbacks posed by the newly elected Reagan administration. Both OSHA and the EPA faced potential attack from antiregulation advocates in the new administration, and some labor and environmental leaders recognized the need for cooperation to defend these key agencies. According to D'Ambrosio,

The fact is that I think that labor unions . . . had come to the conclusion . . . that there were a lot of things that we wanted to do . . . related to labor legislation . . . that we couldn't do by ourselves. We had to go out and start building coalitions with other groups, whether they were environmental groups or religious groups or what have you, in order to help us get things done. The general conclusion was we pretty much had to work together to accomplish what we wanted to do on both sides. (personal communication, 1998)

At the national meeting, labor and environmental leaders agreed to create a nationwide coalition to address federal legislation of interest to both groups. They also decided to build statewide networks to bolster their national efforts and to address similar issues on the state level. D'Ambrosio was among those labor leaders who saw the strategic advantage of coordinating efforts with the environmental community. The diminishing strength of the labor movement necessitated coalition work. D'Ambrosio and other labor and environmental participants from Wisconsin agreed to pull together a meeting in their state. Later in 1981 a meeting was held among labor and environmental leaders in Wisconsin, and the Wisconsin OSHA-Environmental Network (later the Wisconsin Labor-Environmental Network, or WLEN) was founded, with D'Ambrosio as the cochair representing labor, a position he would hold for ten years, until his retirement in 1991.

Prior to the founding of the network, relations between the labor and environmental communities in the state had been tense. Efforts on the part of environmentalists to create deposit legislation for beverage containers had met with fierce opposition from steelworkers in the can-making industry as well as from the unions representing workers in Wisconsin's substantial brewing industry. But the network did a great deal to alleviate these tensions. It initially avoided tackling contentious issues, so that the two sides could first build rapport by addressing issues of mutual concern. Gains made through instrumental cooperation and compromise were encouraging to both sides. According to D'Ambrosio,

As we worked together and got to know each other, there were a lot of things that we thought we wouldn't get along on that turned out not to be so. We agreed on a lot more things than we ever disagreed on. What we tried to do was concentrate on those areas where we could reach agreement. . . . And you'd be surprised how effective this group could be. You walk into a legislator's office . . . and let's say you're a labor leader and you're trying to get this legislator interested in some piece of labor legislation. You walk in there and you've got an environmentalist sitting with you and it makes a big impression. And vice versa. If environmental people want to get something done and they can walk into a politician's office and they've got a labor leader with them, I think it carries a certain amount of weight. These congressmen or congresswomen were glad to see that . . . everybody in the labor movement wasn't against it. It was very effective. (personal communication, 1998)

But over time the instrumental gains to be had from cooperation grew into something more. D'Ambrosio talks about the learning process that network participation meant for him: "I had a lot to learn about the environmental movement. . . . I got a good education, frankly, from meeting with these environmental people and being involved in a lot of these issues, which I either didn't think about before or didn't get involved in personally" (personal communication, 1998).

The heightened consciousness regarding environmental concerns that resulted from his network participation led D'Ambrosio to promote union support of measures that went beyond labor's traditional workplace focus, including support for a community right-to-know measure designed to give community members access to information regarding toxic substances being used in local plants. According to D'Ambrosio, the union supported the measure for a number of reasons:

We definitely wanted to try and help the environmentalists, but the community right-to-know legislation had an effect on communities throughout the state. And you're not just a union member, you're also a citizen and somebody who lives and works in [the community]. And the community right-to-know law could affect you as an individual citizen. . . . There are . . . environmental issues that affect union members as citizens in their own communities. Your members breathe the air in their communities, they swim in lakes and streams, they fish, they hunt, they do all these things . . . so they're affected by environmental legislation whether they're union members or not.

Thus far this book has focused on the way in which external conditions and internal organizational constraints inhibit or facilitate intermovement cooperation between labor unions and environmental SMOs. Prior chapters have demonstrated that various political and economic factors can influence labor-environmental relations and that organizational range indicates which groups are likely to engage in cooperative efforts and which ones are inhibited from doing so. However, it has also been suggested in previous chapters that organizational range is not static. There is evidence that some unions and environmental organizations are currently undergoing an expansion of organizational range and incorporating new issues into their agendas in ways that increase the probability of intermovement cooperation. D'Ambrosio's story provides an example of how this can occur.

Chapter 5 introduced three types of coordinated action in which groups can engage: instrumental cooperation, in which groups work together to promote a single policy that achieves distinct organizational goals; compromise cooperation, in which partial concessions allow for concerted effort to achieve more limited or modified goals; and finally, enlightened cooperation. The support for certain environmental measures offered by D'Ambrosio and the Allied Industrial Workers represents the third form of cooperation. The first two types of cooperation allow organizations to maintain their focus on core goals and to modify only their strategies to improve their likelihood of achieving full or partial victories. To the extent that the unions and environmental groups involved in WLEN sought only to use one another's support strategically to advance their own agendas, their cooperation would be considered of the instrumental or possibly the compromise type.

But D'Ambrosio describes something more. He indicates that new issues were added to the union agenda, issues that he "either didn't think about before or didn't get involved in" before his participation in WLEN. Such enlightened cooperation goes beyond merely strategic considerations and suggests that a new awareness of issues can alter the way in which an organization conceptualizes its members' interests and organizational goals. A significant weakness of much coalition theory is that it focuses exclusively on static interest assessments and strategic calculation and fails to take into consideration the way organizational goals can shift or expand, thus altering the coalition dynamics (Bystydzienski and Schacht 2001).

This chapter focuses on the way organizations "learn" such that they become more likely to engage in intermovement cooperation. This learning can occur when organizations identify a superior way of achieving established goals, but more importantly for our purposes here, it can also involve situations in which organizations come to recognize previously unidentified issues as important and incorporate them into their own efforts. In this study such learning was identified in several cases in which unions or environmental SMOs adopted goals they had never before considered central to their work.

I begin this chapter with a review of the literature on organizational learning, after which I present case study data that suggests that two types of learning can occur, learning through interaction and experiential learning. Next I look at the way organizational leaders attempt to share with their members the new understandings they have developed as a result of this learning. Finally I take up the question of when organizational learning has truly occurred, as opposed to learning that is limited to individuals within an organization.

Organizational Learning

Although much research on organizational learning is based on for-profit firms, many of the general insights developed through this research are applicable to social movement organizations. Many students of organizational learning begin with the perspective that organizations are task-oriented entities that alter their practices to better achieve their goals (Argyris and Schön 1978; Downs and Mohr 1979; Scott 1992). Resource mobilization theory adopts this perspective in relation to social movement organizations (McCarthy and Zald 1977). Yet although organizations adapt for improved effectiveness, there exist structural and normative barriers to organizational learning and the adoption of new practices. Some attribute these barriers to structural "inertia" or "imprinting" that occurs at the founding of the organization (Huber 1991; March and Simon 1958; Stinchcombe 1965). Much research demonstrates that organizations tend to respond to situations using existing repertoires rather than developing novel responses keyed to each particular situation. Once norms for organizational behavior are established, they are adhered to, and this pattern appears resistant to change.

Under stable conditions, organizational inertia will maintain those established practices that are workable. When the organization is faced with changing environmental conditions, however, or unsatisfactory organizational performance that necessitates a response, organizations may adapt. But even when organizations do adopt new practices, they tend to minimize the amount of change they undertake. March and Simon (1958) refer to this as an effort to "satisfice," that is, to seek minimalist solutions to problems that are satisfactory rather than optimal. The decision-making process organizations use to identify the changes necessary to respond to new conditions is still a rational one, but given the complexity of the environment in which they operate, they must employ simplifying methods when they make decisions about change. This tends to minimize the degree of change undertaken.

Research by Chris Argyris and Donald Schön (1978) reinforces this assessment of organizational behavior. Argyris and Schön distinguish between two types of learning that differ in the quality of the change the learning necessitates. The first they refer to as "single-loop" learning, which is directed primarily toward error detection and correction. Single-loop learning, according to Argyris and Schön, "is concerned primarily with effectiveness—that is, with how best to achieve existing goals and objectives and how best to keep organizational performance within the range specified by existing norms" (21). Although such a minimalist strategy may be good for addressing certain problems, in other instances, more fundamental change may be needed: "In some cases . . . error correction requires an organizational learning cycle in which organizational norms are themselves modified" (21). This indicates the need for another form of learning, which Argyris and Schön refer to as "double-loop learning." This type of learning requires the organization to adopt new priorities or to restructure its organizational norms.

Although organizations appear to be good at single-loop learning, double-loop learning is rare. Argyris and Schön attribute this rarity in part to the individuals who make up organizations, their limited understanding of the larger processes that exist outside of their individual area of responsibility, and a lack of communication among different organizational actors. These factors prevent organizational actors from identifying conditions requiring a major reassessment and significant change in organizational practices. When problems are encountered, individuals within the organization may adjust their behavior accordingly, but these individual changes do not lead to or result in fundamental changes for the organization overall. It may even be said that when organizational change is limited to the individual level, no *organizational* learning has taken place at all. Organizations are made up of individuals, yet organizations are more than a mere collection of individuals. An individual in an organization may identify a problem and change their behavior within the organization to correct for it; we cannot say, however, in such a circumstance that organizational learning has occurred unless that change is institutionalized and carried on beyond that individual's tenure.

Single-loop organizational learning is common, as new practices are simply incorporated into the role of an individual's position within an organization. The larger question of concern here, however, is when and how organizations make more fundamental changes. Unions, environmental groups, and other SMOs, like all organizations, constantly modify practices as they encounter difficulties and identify better ways to achieve goals. Coalition strategies, in themselves, are an innovation that many groups adopt as a way to make achievement of their goals more likely. Political organizations typically recognize the benefits of coordinating with others to achieve common goals. But engaging in instrumental cooperation of this sort does not signify a significant change in the organization. It is merely the adoption of a new strategy to achieve existing goals. For instrumental coalitions to become more long-standing alliances among social movement organizations, a more profound form of organizational learning must take place. Thus the key questions are: when are the core goals of an organization modified, and who makes this determination?

When considering changes in more fundamental goals and norms, as opposed to routine tasks, decisions tend to be made at the higher levels of the organization. Political entrepreneurs who found movement organizations define the core goals around which these organizations are centered. Identifying goals involves making basic value judgments about what is good and what the organization should pursue. The goals thus identified and their underlying values then structure decision making at all other levels of the organization as members identify the best ways to achieve the goals that have been established. Thus founders or subsequent leaders high in the organizational hierarchy make decisions that include greater value content, whereas the decisions left to those lower in the hierarchy have a higher factual component, based ultimately on achieving the value-based goals determined by those who set the general direction for the organization (Simon 1976).

Granted, voluntary or democratic organizations generally provide checks within the organization on the decisions made at higher levels and feedback loops that influence the direction in which the organization will move. These forces play an important role in shaping the behavior of social movement organizations, as discussed in chapter 5. As indicated in that chapter, organizational constraints tend to prevent movement leaders from deviating in any significant way from established practices. Although grassroots pressure can on some occasions force change within an organization, more often the assumed desire of a non-rebellious membership to maintain the status quo will act as a constraint on movement leaders concerned about maintaining the organization. When an organization's members are perceived to be satisfied, its leaders are disinclined to pursue a new direction, since the membership has been built on previous and current practices and a set range of issues. But in some cases organizations *do* make significant changes in their goals or practices. There may be pressures from lower levels within an organization working for or against this degree of change, but ultimately those responsible for running the organization are the point of decision making regarding whether significant organizational change is implemented.

The question then is, what motivates organizational leaders to alter an organization's goals or strategies in fundamental ways? Some have suggested that two basic types of learning occur in organizations: one through a process of trial and error and the other through the transfer of knowledge from other organizations (Levitt and March 1988). These two learning processes can lead to both simple error correction and more elemental transformations, although typically the latter must be inspired by external crises (Argyris and Schön 1978; Staggenborg 1986). A major setback or a significantly threatening change in the organization's external environment such that traditional strategies are no longer effective can lead to a reassessment of the organization's basic practices and experimentation with new ways (Banks 1990; Simmons 1990). This is trial-and-error or "experiential" learning. Organizations faced with major setbacks or other organizational threats are more willing to try new things or to tap into their reservoir of experiences to identify strategies that are likely to work. New strategies are adopted (or old ones found to be successful are rediscovered) and those that do not yield the desired results are abandoned. D'Ambrosio indicates that labor-environmental coalitions themselves were a strategy born of crisis. Decline in union density and the realization among labor unions that "there were a lot of things that we wanted to do . . . that we couldn't do by ourselves" led these unions to seek out allies among environmetal organizations.

The second form of organizational learning, learning from other organizations, requires some level of contact with external sources of knowledge. In this type of learning, organizations adopt new practices based on newly acquired information or understandings derived from others' experiences. Contact with external sources of information can be direct or indirect, although typically direct interaction is more likely to bring about the fundamental changes that are of interest here. It is easy to see how these two types of learning, experiential and learning from others, might interact if an organization's ongoing failures or a major crisis facilitates contact with other organizations, which then supplement the organization's experiential learning by providing new strategic insights. Examining the practices of labor unions and environmental-movement organizations reveals both experiential learning and the adoption of knowledge from external sources, namely, one another.

Experiential Learning by Unions and Environmental SMOs

Many of the broad shifts in labor-environmental relations at the national level can be interpreted in the context of organizational learning. For example, the support the labor unions lent to the budding environmental movement and other SMOs in the 1960s can be seen as an innovative effort by the declining labor movement to foster a broader counterweight to growing corporate power. Labor unions' withdrawal of this support in the 1970s derived from the realization on their part that environmental protection might weaken the economy or threaten jobs. Their subsequent building of alliances with environmental organizations in opposition to deregulation and globalization suggests yet another reassessment of when cooperation should be considered advantageous. Each shift in labor unions' attitudes toward environmental organizations (and vice versa) can be taken as an expression of organizational learning on the part of unions and environmentalists as they confront changes in the external environment. They experiment with different relationships and try their hands at working on different issues. Their experiences and interactions facilitate new understandings about their roles, the appropriate positions they should take on issues, and what their relations should be with one another. By applying organizational learning theory, we can interpret the learning process that has led to these fundamental shifts in strategy. The case studies and interviews analyzed here provide more direct evidence of the learning that has taken place among movement leaders, and how that learning has led to decisions regarding the development of new organizational strategies and alliances.

A number of labor and environmental leaders interviewed in the research for this book identified experiences that facilitated a reevaluation of their position on various economic and environmental issues. For example, in Maine, paper companies long had the strong support of workers and paper industry communities. The jobs provided by paper facilities offered wages far in excess of others in these small, relatively isolated communities, earning for these firms the support and loyalty of workers and residents. For many towns in rural Maine, the paper industry was *the* economic foundation that allowed their continued existence. Up until the 1980s industry leaders could count on the support of these communities and of the labor unions when plants were confronted with regulatory or financial challenges. Layoffs and a number of bitter strikes during the 1980s, however, led labor leaders to reevaluate their traditional support for the industry in a range of areas, including environmental protection. John Dieffenbacher-Krall, executive director of the Maine People's Alliance, a grassroots citizens' organization that has facilitated intermovement cooperation, identified a substantial change in labor and community attitudes toward the paper mills during the 1980s:

> It was well understood that the workers, as long as the mills were providing good jobs . . . had their loyalty. If anyone proposed to screw around the paper mill, they were going to have all the workers and the community opposed. . . . But that began changing in the 1980s. The mills started doing a lot of layoffs, and the communities are not so quick to defend them now. For instance, the town of Millinocket decided to raise its mill rate for local property taxes, and Great Northern Paper is objecting: "This is going to cost us $400,000!" [In the past], if the company had made a statement like that, [the community] would have been instantly scrambling to reduce that property tax to accommodate. This time

they said "Screw you. You talk out of both sides of your mouth. On the one hand, you say you want skilled workers, and yet you don't want to pay for education." I think that historical relationship of blind loyalty is starting to deteriorate. (personal communication, May 19, 1999)

This change in attitudes toward Maine's paper industry is clearly evident among labor leaders as far back as the late 1980s, when they took the unprecedented step of supporting environmental measures in opposition to industry wishes. According to Dieffenbacher-Krall,

The culmination of cooperation between the environmental community and labor community was in 1989–90. . . . That legislative session in 1990 passed two major environmental bills with active support from the AFL-CIO. . . . One was a toxic-use reduction law, which was a revolutionary change in the way we deal with toxic chemicals, a preventive approach instead of an after-the-fact approach. The other was a bill known as "Color, Odor, Foam," which would impose standards on how much [waste] is discharged into the rivers. It was really a backdoor way to get at toxic chemicals at paper mills, especially dioxin. And in previous years, paper mill management would just have to call on local union presidents and say, "This is bad." But instead the labor folks said: "We've done your bidding in the past, and we've still had the layoffs and now we are stuck with all the toxic chemicals and the pollution and the sick people. Well, we are not buying it." (personal communication, May 19, 1999)

The labor unions' reevaluation of environmental concerns was also facilitated by strikes at two paper mills, one in Rumford and one in Jay. Striking workers in these towns adopted a range of strategies to challenge the paper mills, including efforts to impose tighter environmental controls. An accidental toxic-chemical release by replacement workers in one of the striking plants added to concerns about environmental contamination caused by the mills. Ned McCann, secretary-treasurer of the Maine AFL-CIO, describes the federation's accumulation of knowledge about the environmental implications of mill practices:

What had been known anecdotally in Rumford and Jay about chemical releases and health issues became more documented, and people studied the effects of chemical and industrial releases and high cancer rates in those areas. So people started identifying health issues and environmental issues as being really the same, and that kind of laid the groundwork for the toxics use effort in which unions joined with environmentalists. (personal communication, May 20, 1999)

These experiences enabled union leaders to identify the relevance of environmental matters to their own work. They began to view the environmental issues as "being really the same" as the health issues that constitute a standard element of union concern. These experiences also led union leaders to question their relationship with employers on issues not directly related to the workplace. Although there is always some level of conflict between unions and employers over issues related directly to work, as indicated by those interviewed in Maine, workers had generally sided with employers when faced with external threats to the industry. It is around these issues that organizational learning occurred. As suggested by organizational-learning theory, this fundamental transformation was inspired by external crisis, in this case layoffs and two bitter strikes.

Organizational learning of this sort was evident in other cases as well. In Washington leaders in building-trades unions identified a growing abuse of prevailing wage rules and an increasing use of nonunion labor in construction projects. Threatened by declining membership, the building trades sought new strategies for confronting nonunion contractors. The unions created an organization called Rebound to monitor worksites for compliance with not only labor law, but environmental law as well. In this way environmental issues were used as an additional means to pressure targeted employers. Employing this strategy created some apprehension, because, in effect, by doing so the union could potentially be reducing the total amount of work available to construction workers, while at the same time drawing attention to the environmental threat posed by development in general. However, this innovation in union strategy came to be accepted as legitimate and ultimately reinforced new thinking about the role of the union in non-work-related contexts. Ron Judd of the King County Labor Council in Washington describes this transition:

We have been known to actually go out and stop projects because of their environmentally destructive nature. The first couple of those that we stopped, when I was with the building trades, that created some debate, some controversy. It was change and change is sometimes stressful. A lot of people didn't know if they wanted to go down this road or not. But after the first couple, and after they realized that the world wasn't going to end and that it actually makes sense and that they are actually doing something that shows that they care as much about the community as their job, [the practice was accepted]. And in fact it was amazing some of the editorials and some of the embracing that the community did on a couple of projects that we opposed. (personal communication, March 29, 1999)

Trial-and-error experiential learning is clearly evident in this case. Union leaders tried taking a stand on environmental issues and learned that they could do so without risking substantial opposition from the rank-and-file membership. There was some resistance and, as discussed in the previous chapter, it is this kind of pressure that maintains a limited organizational range, prevents strategic innovation, and inhibits coalition work. But in this case advocates for change prevailed. Although using environmental regulation as a pressure tactic was originally a defensive strategy used in response to the growing crisis of membership decline, its broader implications should not be dismissed. This is more than simple error correction and better represents double-loop learning, whereby basic organizational goals are altered. As suggested by Judd, the notion of labor action on behalf of the community, as opposed to just focusing on members' work-related interests, was a significant departure from tradition.

In the Maine and Washington cases as well as others, movement actors facing an external crisis began experimenting with new approaches. Unions, originally motivated by a desire to defend jobs and other workplace interests, tried innovative strategies for defending those interests that included work on environmental issues. Success with these strategies led to their incorporation into the union's political repertoire. But more than that, the successful implementation of strategies involving the environment facilitated the development of new concerns and goals on the part of unions. Threats posed to the environment and the community became issues that some unions addressed routinely, rather than just as a means of accomplishing other goals.

Some evidence suggests that a broader learning process is taking place within the union movement in the United States. The thirty-year corporate-government assault on organized labor that began with the recession of the early 1970s has left union leaders groping for effective strategies to counter the decline in union membership and influence (Brecher and Costello 1990). Although organizational inertia can explain unions' continued use of the same failed strategies into the 1990s, new leadership at the national level is now implementing innovations in strategy that were developed by some internationals several years ago. These innovations include recognition that the social contract of the postwar years is dead and that a more militant approach is needed to confront hostile employers. This reevaluation has also led to the search for new allies or reacquaintance with old ones, including members of the environmental movement. This process has been made easier by the wide diffusion of environmental thinking over the last thirty years and the growing recognition within some sectors of organized labor that broader concerns may rightfully be included on the union agenda. Thus, in addition to the specific instances of experiential learning identified in the examples discussed above, a larger process of learning and the implementation of that learning is currently underway within the labor movement.

As suggested in chapter 5, there is some evidence of a corresponding transformation within the environmental movement in terms of the incorporation of more social issues into the agenda of some organizations. Indications of experiential learning leading to this kind of transformation were evident among some of the organizations included in the research conducted for this book. Many environmental leaders interviewed noted that economic issues presented serious difficulties for them. The opponents they faced were often successful at framing environmental-policy debates as one of jobs versus the environment. High-profile disputes such as those surrounding the protection of old-growth forests in the Pacific Northwest indicated that the economic repercussions of environmental policy had to be given greater consideration than the environmental organizations had been giving them. In response environmental leaders began to incorporate more economic analyses into the programs they proposed or advocated. According to a staff member at New Jersey Audubon,

Economics is always there as an issue to be dealt with in the defensive sense. In the history of the environmental movement in New Jersey, it's there as a defensive question because it's almost a reflexive action on the part of the corporate world or any of the regulated parties, whether it's about land use, or water controls or toxic controls, to yell "You are taking away jobs!" The environmental community has had to come up with answers to that, which vary upon the structural situation. Unlike the debates of the 70s, which were "growth versus no growth," that's pretty much evaporated in the thinking of almost all land use groups. We always emphasize in public now, when we are going for tough land use restrictions, that we are not trying to control the level of economic output or construction.

The shift of position on the part of environmentalists in relation to economic issues does constitute organizational learning, but as indicated in the description above, the incorporation of economic issues into the agendas of environmental organizations was, for some of those organizations, simply part of a defensive strategy. In their quest for popular support and perhaps to thwart their opponents, they had to develop arguments to counter the claims that implementation of environmental policies meant job loss. Thus, at least for some environmental organizations, the injection of economic analysis did not indicate that economics was a new concern that would be fully incorporated into their organizational agendas. When asked about job loss issues, several environmentalists interviewed did not fully engage the economic concerns raised by the implementation of policies aimed at safeguarding the environment. Many simply cited research demonstrating that environmental protection is not associated with job loss and that environmental policy results in a net increase in employment, and several used this finding to dismiss the job loss concerns raised by labor leaders. But many did not address the more specific challenges raised by the implementation of environmental policies, such as the fate of those workers who do lose their jobs or the inferior wage levels offered in the "clean" employment sectors that come to replace high-wage manufacturing jobs.

Although the incorporation of more economic analysis into their organizational agendas does represent some change on the part of environmental-movement actors, this alone does not necessarily indicate that they are reevaluating basic values. Rather, they are using new tactics to achieve previously established goals. It is understandable that environmental organizations find it difficult to make more fundamental changes in their agendas. As discussed in chapter 5, environmental and other voluntary organizations are less capable than labor unions of altering or expanding their organizational range to incorporate new values and goals. The members of a voluntary organization may withdraw from participation in the organization if they believe the organization has changed course or that it no longer effectively represents the concerns that led them to join initially. Leaders of these organizations are aware of the risks they face and the need to maintain the focus and image of their organizations if they are to retain their members.

There is still substantial evidence, however, to indicate that more fundamental transformations are possible, even within voluntary SMOs. Although experience facilitated the adoption of economic analysis as a strategy, those environmentalists interviewed in this research were more likely to report the full adoption of job- or labor-related issues if they had engaged in a process of interaction with the labor community. This issue will be taken up in more detail in the following section.

Learning through Labor-Environmental Interaction

Generally organizations are resistant to change. Movement organizations faced with some kind of crisis or significant problem, however, may adopt novel strategies or experiment with the adoption of new goals in response. Organizational learning can occur as a result of this experience. In the cases discussed in the previous section, labor movement leaders learned that employers to whom they lent support on certain issues could not be counted on for fair treatment in other realms. They learned of threats to their interests to which they had not previously given significant consideration. They learned that the adoption of additional goals outside of their traditional area of concern did not generate the expected opposition among the membership and that action to achieve those goals can win praise from other potential allies. Each of these insights generated additional opportunity for labor to work with the environmental community. But each of the cases discussed above involved organizational leaders' experimenting with new practices and learning from their experiences.

As noted in a previous section of this chapter, there is another basic type of learning in addition to the experiential type: that in which an organization adopts new practices based on knowledge gathered from others rather than from its own experience. This type of learning obviously requires some form of contact with other organizations, such as that which D'Ambrosio had through the Wisconsin Labor-Environmental Network. This contact does not, however, necessarily have to be direct interorganizational contact. Knowledge can be attained from personnel who transfer from one organization to another or through written material or hired consultants who bring in knowledge from outside the organization. But the learning that is of most interest here, and for which there is a great deal of case study evidence, is that learning that occurs as a result of direct organizational interaction. As discussed earlier, we are initially interested in examining learning specifically as it occurs among organizational leaders and strategists. Later in the chapter I will take up the question of how learning is diffused through organizations.

Labor union and environmental leaders can come into contact with one another under a variety of circumstances. Sporadic contact occurs between lobbyists from both types of organizations, as they share a common work environment. Organizational leaders also report that they, on occasion, encounter one another at public events. But these incidental encounters, in themselves, rarely offer opportunities for information sharing in a way that would allow for a basic readjustment of organizational values or goals. That kind of sharing requires more concerted interaction, which occurs only under certain circumstances. In the next chapter I will analyze the circumstances that lead to such contact. In this chapter the focus is on what happens in the course of interaction among leaders of organizations in the labor and environmental sectors.

Some labor and environmental leaders describe a process, similar to D'Ambrosio's experience, of learning resulting from routine contact with members of one another's organizations. Of course, whether such learning takes place depends a great deal on the context of that interaction. Some contact may well generate greater conflict between organizations (Marwell and Schmitt 1975). But many unions and environmental organizations have been brought together in the context of resolving differences and forging cooperative ties. Again, this type of contact and interaction typically results from the existence of a significant threat to both organizations or a failure in the pursuit of their major goals. When organizational contact occurs under these circumstances, at least, it is evident that significant organizational learning may occur.

An initial step toward the more fundamental value and goal transformations that eventually occur as organizations interact is the overcoming of stereotypes about one another and the building of trust (Bacharach and Lawler 1980; Nissen 1995; Rose 2000; Starr 2001). Since unionists and environmentalists typically have little formal contact with one another, each is left to rely on media accounts and popular myths for an understanding of the other and the movement they represent. The distorted images that result from lack of contact are overcome as face-to-face interaction proceeds. Each side begins to recognize that the other does not hold a myopic focus limited to its own immediate interests. Labor participants develop an understanding that environmentalists are not fanatical tree huggers bent on destroying jobs, and environmentalists come to appreciate that workers are not ignorant tools of their employers or greedy opportunists, but that they, as individuals, care about the community outside of their workplace.

Union and environmental leaders who have successfully built close ties with one another recognize the need to overcome the stereotypes each side has developed as a result of the lack of interaction between them and to demystify one another's movements. The Labor Institute, an organization affiliated with the former Oil, Chemical, and Atomic Workers International Union, has created a formal training program to try to build labor-environmental solidarity around the nation. Training administered through this program creates a forum for interaction between union and environmental leaders, and part of the program is designed specifically to familiarize each side with what the other actually does. Amy Goldsmith, executive director of the New Jersey Environmental Federation, has worked as a trainer for the Labor Institute program. She notes that the Labor Institute

would start working with those trainers and doing sessions around the country, partly with community environmental people, partly with workers. They would have one worker trainer and one environmental trainer (which I was) and start getting people to engage in dialog to demystify who we were, meaning the environmentalists. Who were the environmentalists? Who were the community people? Were they really the enemy? Or was there actually some common ground? Basically the labor people thought we were all Greenpeace who climb the stacks and want to shut them down. We would go through this training program together and people would realize that we don't want to shut them down. We want it to be safer and want them to be protected just like they do. So people would leave realizing that we have more in common. (personal communication, May 26, 1999)

D'Ambrosio, reporting on his experience with the Wisconsin Labor-Environmental Network, remarks on the importance of the familiarization process: "When you got to know people better ... who maybe in the past you looked at skeptically, you found out that you could get along and have trust and confidence in them" (personal communication, 1998).

D'Ambrosio and others also stress the importance of interaction outside of negotiation and political strategizing. Informal contact is particularly important for the trust-building element of coalition work. Anne Rabe of the Citizens' Environmental Coalition is one of the central figures in the labor-environmental network in New York. According to Rabe,

It really helped to socialize. To socialize between labor and environmental groups was always key from the beginning, so we always made time for socializing at [our] conferences. The steering committee would go out to dinner and that sort of thing. So you can get to know people as people, and try to break some of the stereotypes that environmentalists have about labor, and labor has about us "tree huggers." . . . That kind of socializing and networking was really helpful. (personal communication, May 11, 1999)

Some researchers have found that informal encounters between coalition participants can also serve as important solidary incentives for coalition participation (Hathaway and Meyer 1997). Labor and environmental leaders involved in coalition work in Wisconsin, Maine, and New York reported in interviews not only that their cooperative efforts had been effective, but also that they enjoyed their coalition work very much. According to Ken Germanson of the Wisconsin Labor-Environmental Network, participants in the network "enjoyed being with each other. I think that that's all part of networking and coalition building is to begin to enjoy a relationship, so that you look forward to coming to a meeting, if for nothing else then to share some good times with some folks. And I think that's an important part of political action" (personal communication, Oct. 10, 1996). Experiences as simple as meeting with other people and breaking out of standard routines provide important incentives for coalition activity.

Direct encounters not only help to build trust among coalition participants, but they enable each side to understand more about the movement represented by the other side and the organizations with whom they are dealing. Both sides learn one another's basic organizational norms and in this way avoid confusion or missteps that can send the wrong signals. For example, union leaders come to appreciate the diversity represented among the various actors in the environmental movement and begin to distinguish between the groups that are more likely to be open to the labor cause and those that can be expected to be less responsive to labor concerns. Dan Becker of the Sierra Club reports having to correct a labor speaker repeatedly at a meeting in which the union leader kept referring to "environmentalists" having taken a certain antilabor position. Becker interjected each time to reinforce the fact that the Sierra Club had never taken that position, although some other environmental organizations had.

Carl Pope, the executive director of the Sierra Club, explained to labor leaders in their own terms the club's inability to control every environmental activist. When union leaders complained about a Sierra Club activist in Ohio who adopted a position they opposed, Pope explained that, like unions, environmental organizations have "wildcat strikes" referring to instances in which local union workers go on strike without the authorization of the union leadership. This kind of exchange serves an educative function, in this case making unionists more aware of important divisions within the more decentralized environmental movement. The deeper understanding of one another's movements that results from such interaction contributes to the trust-building process among coalition partners.

Overcoming stereotypes and building trust are only the first steps in the organizational learning process that concerns us in this chapter. Once the foundation provided by the eradication of stereotypes and the fostering of trust has been laid, coalition participants' understanding of the issues raised by their coalition partners expands significantly. This expansion is important, because movement actors working in a particular movement sector are often unfamiliar with the work of other groups in a different sector. As described in the last chapter, political structures in the United States foster the formation of narrow, single-issue interest groups. This can leave the leaders of a particular movement relatively isolated from actors in other movements. Coalition activity or other forms of routine interaction can provide an educational forum in which participants learn about issues beyond their own limited focus (Brecher and Costello 1990; Childs 1990; Nissen 1995; Richards 1990; Rose 2000). As D'Ambrosio's earlier comments illustrate, this can lead to an internalization by one movement of the values associated with the goals of other movements. Ed Ruff of the New York State AFL-CIO described a similar experience with the Labor and Environment Network in New York: "The closer we worked [together], and the more labor people we brought on board to start looking at the big picture of not only the workplace as the environment but the community, we learned that there were more and more environmental bills that [we could work on]. . . . As far as the coalition is concerned, there was definitely new thinking for the people that got involved" (personal communication, May 12, 1999). This "new thinking" about "the big picture" also developed among environmental participants. Whereas many environmentalists dismiss job loss fears out of hand as industry manipulation, direct contact with workers' representatives has allowed some to recognize the legitimate concerns that underlie some worker opposition to environmental measures. As the director of the Sierra Club's Global Warming and Energy Program, Dan Becker was very involved in the Climate Change Working Group organized by Jane Perkins at the AFL-CIO. He describes his experience as he became more involved with labor-environmental negotiations over greenhouse gas emissions:

Until I started negotiating with the folks at the AFL, I unquestionably bought the win-win "energy efficiency has no losers" rap. . . . But it finally dawned on me in the course of these discussions that if you are burning less energy because you are using it more efficiently, then whoever digs out the energy stuff is somehow going to have the potential of suffering. . . . Clearly as we move forward with energy efficiency and other solutions to global warming, there will be some people who will pay a price. Some may lose their jobs. Some *will* lose their jobs. . . . It should be part of the environmental approach that those workers not pay a price for society's benefit and the planet's benefit. We need to provide them a just transition through that difficult time to a clean energy future. (personal communication, June 2, 1999)

That concerns about protecting union jobs and providing assistance to displaced workers should become part of the Sierra Club agenda demonstrates a profound form of organizational learning for an environmental group.

Of course, full transformation in the thinking of movement actors does not occur in every instance of organizational interaction. On the contrary, coalition participants often refer to the need to "agree to disagree" on certain issues (Altemose and McCarty 2001; Brecher and Costello 1990; Cullen 2001; Richards 1990). For reasons discussed in chapter 5, some issues are considered too central to the core mission of these distinct sectors to allow for a radical reevaluation of goals. This is especially true of environmental groups, who, as noted above and in previous chapters, must rely on the voluntary support of members. Thus, successful coalitions typically allow for some issues to remain decidedly off limits. Coalition partners then act independently of one another on those issues in accordance with their own organizational needs. On other issues, however, even when full transformations of organizational attitudes and goals do not occur, the learning associated with organizational interaction at least facilitates a greater understanding of the issues and creates the potential for more in the way of compromise cooperation.

Educating the Membership and Others

Learning on the part of an organization's leaders is crucial if new goals are to be incorporated into the organization's agenda. As the key decision makers, its leaders are in a position to decide whether or not a particular issue is given real consideration in the day-to-day operations of the group. It may not be feasible, however, for an organization's leaders to decide independently to incorporate previously unrelated issues into the organizational agenda without first winning the consent of the membership. In some instances it may be possible to do this, as rank-and-file members may be unaware of the specific actions being taken on behalf of the group (Moe 1980). Yet in many of the cases examined in this book, leaders took it upon themselves to try to educate the members of their organizations about the importance of new issues they wished to pursue as part of the organization's agenda.

One of the main means of contact between professional environmental organizations, like the Sierra Club, and their members is through their publications. Becker notes that through the Climate Change Working Group's discussions with labor leaders, he and other Sierra Club leaders began to internalize new ideas about the need to incorporate worker issues into environmental policies. But they were also very aware that their members, recruited through environmental appeals, did not necessarily share these ideas and that the abrupt adoption of a new approach that took this need into account could prove disruptive to the organization. In order to address this concern, they engaged in concerted efforts to raise the awareness of their members of the wisdom and necessity of making worker concerns part of the group's agenda. These efforts involved coverage in the published materials sent to members, including a feature article by David Moberg in its membership magazine, *Sierra*, entitled "Greens and Labor." According to Becker,

As these negotiations have progressed, one of the things that I talked with the executive director about is that we need to prepare our members for a point that is rapidly arriving when, if we are successful, prolabor positions will be part of our agenda. As part of a just transition, we will favor things that we have not been in favor of before. For example, the Abandoned Mine Reclamation Act provides hundreds of millions of dollars to repair abandoned mines in Appalachia, something we never really worked with. Now we are going to work on it, and we are going to hope that it provides jobs for coal miners who have lost their mining jobs. And those are going to have to be union jobs. It will be an environmental position that the Sierra Club advocates. As we advocate more money for solar rooftops, renewable energy, those should be union jobs. Those should be represented workers putting in those solar panels, and those are good building trades jobs. So part of the purpose of the Moberg article was to begin educating our members that we have a shared stake in making progress with labor, that it isn't limited just to NAFTA, but that it extends to global warming and other issues, and that things that have not been traditionally environmental issues, like making sure that a job is a union job, or guaranteeing the right to organize—that these are now Sierra Club positions. We can't just burst on the scene and say: "OK guys, this is our new position." We have 600,000 members, we have to begin to educate them, and that's one of the things you do. (personal communication, June 2, 1999)

Such an effort by environmental-movement leaders to educate members is evident in a number of other cases. The Labor Institute program mentioned earlier in the chapter is a more systematic attempt on the part of both labor and environmental leaders to educate rank-and-file workers and grassroots environmentalists so that they appreciate the concerns of the other side in labor-environmental issues. As part of the program, A Just Transition for Jobs and the Environment, labor and environmental trainers conduct a series of workshops with local workers and members of the environmental community. The workshops take participants through a series of discussions, some conducted with workers and environmentalists separately, some conducted with them together, in which sentiments about the environment and the economy are analyzed from various perspectives. Information is provided about the repercussions of environmental destruction in terms of health and the loss of wilderness, as well as about job insecurity, wage concerns, and unemployment.

The Labor Institute program, which was coordinated by officials with the Oil, Chemical, and Atomic Workers International Union, has been conducted in dozens of localities around the nation, primarily with OCAW locals and environmentalists in the same areas. The New York Labor and Environment Network carries out similar efforts through publications and conferences. For example, at one of its annual conferences, it held a workshop for environmentalists entitled "Labor 101: Building Coalitions with Labor Unions." Other sessions at the same conference involved discussions and presentations regarding the importance of one another's issues and the need for each side to incorporate the other's platforms into its agenda.

In each of the cases where labor-environmental cooperation was being carried out, there were corresponding efforts to "move our membership along," as one of the leaders interviewed put it. This practice contradicts the claims of some movement theorists regarding the moderating influence leaders exert on formal organizations, the so-called iron law of oligarchy (Michels 1949). There may be some moderating tendencies among those who head organizations, due to the fact that they develop a more concentrated concern for perpetuating the organization itself (although, as stated before, organizations are necessary to achieve the desired goals around which they are initially built, and thus an organization's survival is not a distinct interest separate from the organization's original goals) (Dreiling 1998; Piven and Cloward 1977). For example, the national leaders of the United Auto Workers have opposed increases in fuel economy standards out of fear of losing jobs and members even though UAW members overwhelmingly support the higher standards (Moberg 2002). This could be interpreted as an oligarchic leadership protecting its own organizational interests at the expense of the broader interests of the organization's members.

However, applying the moderating-leadership thesis in this case is complicated by the fact that supporting higher fuel economy standards involves the introduction of *new* items onto the organizational agenda.

It is not clear whether this should be interpreted as a conservative or a progressive position. In terms of the specific goals originally embraced by these two distinct movements, alliance formation and the expansion of organizational goals, as advocated by movement leaders, could itself be considered a move toward moderation in some sense. For example, environmentalists who incorporate economic or social issues into their analyses of environmental issues may be inclined to accept weaker environmental measures than those who concern themselves only with environmental protection. Yet most would probably consider the broadening of agendas associated with labor-environmental cooperation to be a move toward a more radical transformative program, not a trend toward moderation (Foster 1993).

If we consider the broadening of organizational agendas merging labor and environmental concerns to be a progressive shift, then organizational leaders often play the role of a radicalizing force, contrary to much movement theory. The evidence derived from this study suggests that leaders often advance the expansion of organizational agendas as a result of their experiences with other organizations in the political sphere. This is understandable given their direct encounters with other organizations and the socializing effect that such contact can have. In some cases a grassroots movement may advance organizational changes against the will of conservative leaders, but this is less likely than leader-promoted changes, given the incentives that originally motivated rank-and-file participation in the organization. Individuals at the grassroots level are motivated to join an SMO, at least in part, by a particular grievance or concern. Leaders of the organization share that concern, but because of their direct experience with the political process and encounters with other aggrieved groups, they can develop a broader sense of the need for social change that the grassroots members may lack. As evidenced by the data gathered for this book leaders then act to educate the grass roots, thus spreading a more encompassing analysis of existing social problems. Though often cited as "iron," the law of oligarchy may actually be made of a more flimsy material. Although its effects are evident in some cases, empirical research on the beliefs of leaders relative to members indicates that leaders are often more progressive and militant than their members in at least some ways (Rucht 1999; Sabatier and McLaughlin 1990). The evidence presented in this book suggests that leaders play an important role in broadening organizational agendas to promote more encompassing social changes. At the very least, the iron law needs more critical examination than it has been given up to this point.

Of course, in some instances, innovation at a high level of an organization is introduced from below. National or state organizations with local chapters or other subunits may adopt the best practices developed at the local level. When this has occurred, it has been the local chapters, rather than the national or state organizations, that have undergone the process of experiential learning or learning through interaction with other groups. But the capacity for such local innovation and the propensity for new practices to spread through an organization depend on the relationship between the organization's central body and its affiliates. In organizations with centrally controlled, top-down structures, local chapters may be constrained in their ability to innovate. Such an organizational structure would exemplify the traditional view regarding oligarchy. There may even be struggles for power in organizations with this kind of structure when leaders are resistant to change supported by the grassroots. Some sectors of the labor movement have seen rank-and-file progressive reform movements arise to challenge ossified leaders.[1] In rare instances this can result in a change in leadership, but adroit leaders monitor the work of local branches and incorporate effective new practices and the sentiments of their constituents, which can serve as another basis for broader organizational learning.

Ensuring that rank-and-file members are on board with newly adopted organizational goals is an important task for leaders, but efforts toward this end are not limited to educating the grass roots. Several labor and environmental leaders interviewed who were involved in coalition work described their efforts to bring other organizations within their sector into the coalition. According to the Citizens' Environmental Coalition's Anne Rabe, "Our organization, because we work with so many community groups on the local level, would always try to impress on activists that they need to be sensitive to the workers' concerns if they were attacking a plant. And we were successful in building some local labor-environmental alliances that worked, while in the past there was total polarization between the labor union and the community group" (personal communication, May 11, 1999). Amy Goldsmith of the New Jersey Environmental Federation describes a similar effort to move other organizations: "We are one of few organizations [to work with labor in New Jersey] and we often push other environmental organizations to see the importance of that. . . . Even the more conservative environmental groups are beginning to understand, and it's because we've done some work to educate them. It's not just educating ourselves within the organization, it's trying to educate the environmental community at large" (personal communication, May 26, 1999).

Encouragement from fellow activists within the same sector plays an important role in the further expansion of coalition work. In all of the coalitions examined in the research for this book, environmentalists were assigned the task of outreach to other environmental organizations and union participants were responsible for contacting others in the labor community. Obviously organizations and their members are more likely to know, and to know of, other groups in their own movement than would be those from other movements; thus there are practical reasons for this division of labor, but there are strategic benefits as well. According to Rabe, "We would run into a new labor activist who was very rigid, very suspicious of the environmentalists. . . . We would have the AFL-CIO or the UAW guy go over to them and say, 'Look, these people are okay,' and they would lobby this labor activist to at least extend a hand and listen to what the concerns were of the environmental group in his community" (personal communication, May 11, 1999). Carol Terrell, the Sierra Club's representative in the Wisconsin Labor-Environmental Network, also emphasized the role environmentalists played in integrating other environmentalists into the group, even when that role was not an active one: "There was always somebody new coming on the scene. If they come into the network meetings, and they've never been there before, they look at the interactions [between the environmental and union representatives]. They see that there is trust and sharing of responsibility and information exchange. Then they say, 'It's OK for me to do that,' and they are more willing to start to build a relationship themselves" (personal communication, July 15, 1999). Thus there are actually two forms of learning that occur through interaction, that which occurs as a result of direct interaction between labor and environmental organizations and that in which groups learn from others within their own movement sector about the importance of incorporating other concerns into their efforts.

Although organizational learning can take place in several ways, an assessment of the relative significance of the different forms of learning for establishing intermovement ties suggests that direct labor-environmental contact is most important. In states in which there is an ongoing forum for interaction between labor and environmental organizations, both sides have demonstrated a greater commitment to the issues themselves than in states that lack such an ongoing forum. The degree of this commitment can be seen as existing on a continuum, with those organizations with the least level of contact (and thus commitment) at one end and those with the greatest at the other. For example, in New Jersey, where the labor and environmental communities have little contact, we see sentiments such as those expressed by the New Jersey Audubon staff member quoted earlier in the chapter. As the comments of that staff member show, some learning took place over time regarding New Jersey Audobon's need to incorporate economic issues into its environmental positions, but these issues have been incorporated only in a "defensive," strategic way—the product of experiential learning and pressure from other environmentalists, but without the deeper commitment derived from interorganizational contact. The one environmental group with extensive contact with unions in New Jersey has fully integrated labor issues into its analysis and pushes other environmentalists, with only moderate success, to do the same. In Maine, Wisconsin, and New York, where labor-environmental contact is greatest among the states included in this study, both sides demonstrate a deeper commitment to one another's issues than is found in states with lesser degrees of labor-environmental contact.

In Washington, where contact between labor and environmental organizations is sporadic, it can readily be seen within the environmental community that the agendas of the two types of organizations are not well integrated. An incident in 1999 in which environmentalists called for the shutdown of a coal-fired power plant that the state AFL-CIO was fighting to keep open demonstrated this point. Environmentalists attributed the confrontation to a breakdown in communication between organizations on the two sides. It is probably the lack of routine contact with unions, however, and the labor consciousness that it fosters that led environmentalists to take a position without full consideration of the consequences for employment. Organizational learning can occur in several ways; the indicators, however, are that direct intermovement contact leads to a more substantial integration of the concerns of both movements into one another's goals and agendas.

Organizational Learning or Learning by Individuals within Organizations?

The question arises as to whether the learning that takes place within an organization is rightfully considered *organizational* learning. As discussed earlier, individuals within an organization may identify a problem and change their practices to carry out their duties more successfully. Yet we should consider this form of learning to be organizational in nature only if the new practices are institutionalized and outlive the individual holding a particular position at a particular moment in the organization's history. If the new practices end when the individual who initiates them departs, no organizational learning has transpired. A similar question can be raised about the more significant transformations in values or goals that occur within organizations, especially if the new thinking that results is limited to organizational leaders (Nissen 1995). An organization's leaders are in a position to redirect the actions of the organization, resulting in new practices and the pursuit of new goals; it is unclear in many cases, however, if the new practices and goals will be sustained under successive leadership. The study presented in this book offers no evidence on which a definitive determination can be made about whether values and goals newly adopted by organizations as a result of coalition participation will remain on the agendas of those organizations. I suggest, however, four criteria for determining whether and measuring the extent to which learning has occurred in an organization and whether it will continue to engage in cooperative action.

The first of these criteria is the extent to which the leadership passes on its learning to the membership. As indicated earlier in the chapter, several organizations that had expanded their agendas to include additional issues undertook efforts to educate their membership about this change. Several of these organizations that are now involved in coalition work hold regular conferences focused on forging intermovement cooperation that grassroots members and activists are invited to attend. Such conferences are held in New York and were formerly held in Wisconsin. These conferences feature workshops in which participants are educated about the importance of the issues raised by organizations on the other side of the labor-environmental divide and the need for cooperative efforts. Some organizations also include coverage of coalition activity in their newsletters, as the Sierra Club does in its national membership magazine. Repeated exposure in any forum to the association between the labor and environmental movements and the values they both embrace is likely to have the effect of educating the membership and thus institutionalizing cooperative activity as part of the organization. Failure to educate the membership in this regard and to persuade members of the need for coalition activity on a broadened organizational agenda presents a constant threat of organizational weakening or demise, as members who have not embraced the new issues may be inclined to withdraw from it. Leaders play a key role in coalition activity, but coalition campaigns face a greater risk of failure if the associated organizations are not internally mobilized around the issue on which the coalition has been formed (Nissen 1995). Of course, for the reasons discussed in chapter 5, voluntary organizations run the risk that publicizing their coalition ventures may alienate existing members who are not convinced of their value. Those who joined such organizations because of their own support for particular goals may withdraw from them when they see other issues being interjected into organizational agendas. For this reason it is wise for groups that wish to pursue and sustain coalition activity to include reference to it in their recruitment or initiation materials, so that new recruits recognize that coalition activity is part of the organization they are joining.

This leads to the second criterion for determining the existence of organizational learning and the possibility of sustained coalition action. Organizational learning can be said to have taken place if the new goals are incorporated into the organization's formal platform. As demonstrated in chapter 5, groups that have a broader organizational range are more capable of engaging in coalition activity than those with ranges that are more narrowly defined. Similarly, groups that engage in coalition activity need to formalize their organizational range to include the new issues on which they work as a result of that activity. Acceptable behavior in an organization is indicated not by traditional practices alone, but also by the documentation of those practices (Dery 1998). Incorporating new issues into an organization's platform is a key element of organizational learning.

A third criterion for determining whether organizational learning has occurred is whether the organization officially affiliates with a coalition organization. Coalition organizations themselves vary in the extent to which they are formalized (Brecher and Costello 1990; Mann 1990; Starr 2001; Zald and McCarthy 1980). Some coalition organizations consist simply of very loose, irregular meetings among organizational representatives. Others are formal bodies with independent funding sources, official membership criteria, dues, and officers. These features allow for more successful coalition functioning (Staggenborg 1986), but affiliation with these more formal organizations also means making an institutional commitment on the part of member organizations. Like an official change in the platform of a group, formal affiliation with a coalition organization indicates that the coalition goals are an established element of the group's agenda. Issues around which there is less formalized cooperation are more easily dismissed by new leaders of an organization or by the same leaders at a later date when involvement in the coalition is less strategically convenient. Ad hoc quid pro quo coalition arrangements represent little in terms of actual organizational change.

The fourth and final criterion for determining whether organizational learning has taken place focuses on the key role leaders play in the organization. Since leaders play an important role in determining the direction an organization takes, those who adopt new values and goals in response to coalition participation can sustain coalition participation by ensuring that those values and goals are passed on to subsequent leaders. It is conceivable that even if the membership is not brought on board and even if the new goals and values are not formalized in the organizational platform or through official coalition affiliation, the new practices can be sustained through the leadership alone, provided subsequent leaders embrace them. Organizational learning, by definition, involves the continuation of innovative practices beyond the individual who engaged in the original innovation. Since a group's leaders are in a position to set the course for the group as a whole, if subsequent leaders adopt the practices initiated by their predecessors, then we can say that the organization has learned. Of course, a coalition sustained only through high-level ties will be less effective and more vulnerable to grassroots objection than those that have the active support of members.

Whether an organization demonstrates these four elements of organizational learning—commitment of the organization's membership to new organizational goals, commitment of subsequent leadership to these new goals, incorporation of these new goals into the official organizational agenda, and formal affiliation with a coalition organization—indicates whether organizational learning has taken place. Although the presence of all four of these elements confirm organizational learning, the presence of any one or a combination may be enough to enable the organization to sustain cooperative ties with other organizations. The key question in determining whether organizational learning has taken place is whether the pursuit of new goals outlives the individuals who were instrumental in bringing about the change.

Although it is impossible to make a determination regarding organizational learning among the organizations investigated in the research conducted for this book, some indicators suggest that certain organizations failed to institutionalize the changes that they adopted. For example, the Wisconsin Labor-Environmental Network became dormant in the mid-1990s soon after Ken Germanson, a key organizer for the coalition, retired and the Allied Industrial Workers International Union merged with another union and closed its Milwaukee office. Lacking key personnel and a central coordinating headquarters, the network lapsed, and contact between labor and environmental leaders became sporadic. Because many of the same personnel still head several labor and environmental organizations in the state, they have been able, since the demise of the network, to unite on an ad hoc basis when needed. For example, they were able to unite and diffuse at least one potential jobs-versus-the-environment crisis in the state since the decline of WLEN. Other than the personal contacts that remain from the coalition's active period, however, there is no evidence that sensitivity to one another's issues among the organizations that belonged to the network will continue when the leadership of those organizations changes.

In some other cases, learning has been more institutionalized. The Citizens' Environmental Coalition is the central coordinating body for the Labor and Environment Network in New York, along with a progressive independent labor support organization. CEC has fully incorporated worker issues into its agenda, and its ongoing ties with the state AFL-CIO, even through changes in key personnel, suggest that its commitment to mutual goals is ongoing. The coalition regularly publishes a newsletter and has held conferences consistently since the late 1980s. In Washington, the building-trades unions successfully institutionalized environmental concerns through the creation of an organization called Rebound, which employs lawyers and investigators to pursue violators of environmental and labor laws. This institutionalized incorporation of environmental issues into labor activity suggests that organizational learning has occurred in a profound way.

Conclusion

The cases discussed in this chapter suggest that there is a major weakness in much coalition theory. In traditional coalition theory, organizational goals are usually considered to be static, and cooperative relations are analyzed only in terms of when organizations' goals overlap. Although organizations may be resistant to change, the evidence presented in this chapter demonstrates that goals evolve through a process of organizational learning. This learning occurs in two ways, the first of which is through experience. Movement organizations faced with ongoing failure or a major crisis experiment with new strategies in response. In the process they may identify new and more effective ways of achieving traditional goals, but more importantly, this experimentation can lead them to recognize new goals and to incorporate them into their agendas. This type of transformation often occurs at the top of an organization, where most strategic decision making is carried out, but in other instances grassroots pressure or innovations developed at the local level can diffuse upward.

The second pathway through which organizational learning takes place is not one that occurs through a process of trial and error, but rather one in which interaction with other organizations introduces new ideas. This form of learning accounts for most of the major value transformations that occurred within the movement organizations examined in this chapter. Learning of this type typically begins with trust-building measures and instrumental forms of cooperation. As movement leaders get to know one another as individuals, they abandon the stereotypes that have inhibited positive relations in the past. The more interaction that occurs, the more likely are the participants to incorporate the concerns of one another's organizations into their own. This homogenization of concerns allows still more cooperation and stronger interorganizational bonds.

Organizational learning of this type cannot be considered complete however, unless the new practices and thinking are carried over beyond the individuals directly involved. The interview data presented in this chapter describe a process of dissemination in which leaders shared their newly acquired sentiments with members of their own organizations and others in their sector. If new practices and goals are incorporated into the official agenda of an organization or if its members and subsequent leaders are effectively brought on board with regard to these practices and goals, we can say that organizational learning has truly occurred.

Earlier it was shown that external conditions can facilitate or inhibit coalition activity. Economic and political conditions can create systematic pressure toward cooperation or conflict among unions and environmental SMOs. Yet other evidence suggests that structural factors associated with the external environment in which an organization operates cannot fully explain when coalitions will form and that we must also consider variables at the organizational level. Thus far we have examined organizational range (chapter 5) and the role that it plays in either facilitating or inhibiting coalition activity. In this chapter organizational interaction and experience were identified as key to organizational learning, which then allows for greater intermovement cooperation. In the next chapter we will examine how organizational interaction comes about, thus allowing this learning to occur.

7

"They're Good People; You Should Talk to Them": The Role of Brokers and Bridges

Ron Judd is the executive secretary-treasurer of the King County Labor Council. King County includes Seattle and the surrounding areas in Washington. The labor council coordinates activity among the 175 different labor organizations in the area, which encompass almost 200,000 unionized workers. Judd got his start in the labor movement with the International Brotherhood of Electrical Workers while working in the shipyards of Seattle. He served as a steward and then as a business agent before moving on to head the Building and Construction Trades Council for the region.

But prior to his involvement with the labor movement in Seattle, Judd was in a different line of work: He was an environmental educator and activist. He helped to form an organization in his home state of Kentucky that worked to protect the Red River Gorge. His group engaged in environmental education in addition to organizing outdoor recreational activities. One might think that having environmental sympathies would serve as a hindrance to someone working his way up the labor leadership ladder, especially in the building trades, in which conflicts with environmentalists are not uncommon. But Judd did not hide his environmental inclinations: "I was an environmental activist before I got involved in the labor movement and I still today am very involved. And I make no bones about it. When I ran for this office I ran letting everyone know that I was a strong environmentalist and that that should not be a bad word . . . and that they got what they got if they elected me" (personal communication, March 29, 1999).

The tension between Judd's environmental sensibilities and his job representing the interests of those paid to clear land and construct buildings was put to the test when the Washington legislature was addressing the issue of growth management in the early 1990s. Washington experienced significant population growth during the 1980s, and rapidly increased development was threatening the natural environment and the general character of many local communities. Environmentalists were pushing for strong measures to control growth through new regulations and planning requirements for future development. Developers vehemently opposed these measures and, as is typical in political struggles of this sort, they solicited unions to join them in their efforts to defeat the legislation. According to Judd,

There was huge debate over it. There was almost blood on the floor. I was with the Building and Construction Trades Council, heading up that at the time, and we're very quick to think about the job side.... The developers were really targeting growth management as an antidevelopment strategy, where if growth management passed in Washington state that the ability to develop property, the ability for the economy to grow, would actually cease to exist and tens of thousands of jobs would be lost and construction would go in the toilet. So they were applying the fear factor. (personal communication, March 29, 1999).

Although concerned about the economic interests of his members, Judd did not view growth management as a threat to jobs. After analyzing the proposed measures, he was convinced that they would be good for the state, both economically and environmentally. He then sought to convince his fellow unionists to join with the environmental community in support of the measures. This met with a great deal of resistance, but because of his unique background in both the labor and environmental movements, Judd was able to present environmental concerns in a way that resonated with building-trades workers:

I in fact debated the industry in Spokane about this in front of about 500 building and construction trades delegates at a convention in which we took the position not to support their efforts, but to support growth management. I asked a couple of very simple questions. My questions went something like this: How many of you in this audience have served in an apprenticeship program? Well, virtually everybody's hands went up. And I said, what is one of the most fundamental things you learn as an apprentice? You learn that you don't go out and build a house or a bridge or a high rise or a mall without having a set of blueprints which guide you and ensure that what you are building is sustainable over the long term. I said, so why would we consider building our communities without a blueprint? That's what growth management is all about. It's not about not wanting growth, but about the kind of growth that we want to have. Where is it most appropriate for the growth to go? What are the kind of areas and treasures in our community that we want to protect for ourselves for our kids and their kids? And so why wouldn't we support the concept of developing a blueprint for our communities? They understood that and it really did shift the debate on growth management. (personal communication, March 29, 1999).

Labor's support for growth management in Washington is just one of several efforts that Judd has undertaken to build stronger links between unions and environmentalists. Through his own work with the environmental community he and others have done much to overcome the tensions that existed in the past between unions and environmentalists around issues like nuclear power and the Alaskan oil pipeline. Although their role is often overlooked, individuals like Judd are crucially important for the establishment of positive ties between distinct movement sectors.

Most analysis of coalition formation begins with the assumption that coalition activity occurs when the interests or goals of two or more organizations are in alignment (Zald and McCarthy 1980). Coalition theorists then examine various strategies that would maximize utility for the respective organizations to determine when coalitions will actually form. The last chapter addressed a major shortcoming of this approach: the idea that interests or organizational goals are static and that coalitions emerge only when the fixed objectives of distinct organizations coalesce. Although this kind of instrumental cooperation is most common, other forms of coordinated action can also occur. Organizational learning can result in a shift or expansion of organizational range, allowing for coalition activity when it was originally unexpected. Few would have expected the building trades to support policies that could pose an impediment to development. Yet in Washington the building-trades unions adopted a different approach than the one that would have been expected, one that incorporated concerns that extend beyond the traditional union focus of wages, hours, and working conditions to include broader quality-of-life issues.

As detailed in the last chapter a common stimulus for this kind of expansion of organizational range is direct interaction with other organizations from different movement sectors. Yet the structure of the political system in the United States tends to foster a narrow, single-issue focus among organizations. This narrow focus rarely brings organizational actors from different movement sectors into contact with one another. Each organization develops its own area of expertise and carries out its work in relatively isolated circles of professional researchers, founders, government agencies, and legislative committees. Lobbyists and activists in the organizations examined in the research for this book report only rare occasions in which they encounter one another in hearing rooms or in the halls of the legislature.

The question that then presents itself is, given the importance of cross-movement interaction for organizational learning and coalition formation, how is the context for organizational learning created? Although certain structural pressures may stimulate organizational learning and the adoption of coalition strategies, which can then further expand the range of organizations that participate in the coalition, the key role played by certain individuals in the process should not be ignored. Not only must these individuals establish contact with individuals or groups outside of their central domain, but they must also devise ways to establish common ground with these individuals or groups. Indeed, most examinations of coalition formation fail to consider how contact is initiated, even in instances in which the goals of distinct organizations are in alignment. Many simply assume that shared interests will automatically generate organizational cooperation. The actual processes of making contact and arriving at shared understandings on issues or strategy must be given greater consideration. This chapter is designed to explore these processes.

Three elements were found to be important regarding when and how preliminary contact is made between movement organizations for the purpose of concerted action. The first concerns the actual individuals who reach out across organizational lines, those referred to here as "coalition brokers." A second element concerns the type of organizations that are involved in bringing about interorganizational contact. Although most coalition brokers identified in the research for this book were based in one of the two movements of primary concern, certain types of organizations were found to be more suited than others for building bridges between unions and environmental SMOs, including some third parties that are not among the sectors of focus here. The third element in need of exploration is the issues around which unions and environmental organizations initially come together. Some areas of mutual concern have already been identified in previous chapters, but it was also found that certain issues more often serve as the basis for early intermovement contact than others because of the ease with which they can be framed in ways appealing to both movements. Following a brief review of the relevant literature, this chapter will examine each of these three elements in light of the evidence from the case study research.

Brokers and Bridges: An Overview

A number of scholars have identified the role played by key personnel in bringing movement organizations together for the sake of coalition activity (Brecher and Costello 1990; Bystydzienski and Schacht 2001; Grossman 2001; Loomis 1986; Mann 1990; Rose 2000; Sabatier 1988; Sink 1991). In examining urban policy coalitions, David Sink (1991) breaks down the roles carried out by these "brokers" in the process of coalition formation. The first of these is to identify groups suitable for coalition participation, or stakeholders, around a particular issue. In relation to a particular issue being addressed by an environmental SMO, for example, suitable candidates for coalition partnership may be workers whose health is imperiled by exposure to toxic chemicals on the job, child welfare advocates who are concerned about the susceptibility of children to environmental contaminants, or homeowners worried about property values. A coalition broker identifies those organizations that may have at least some interest in a particular issue of focus.

Once potential coalition partners have been identified, the broker must conduct outreach to them and provide the motivation for coalition participation. This coalition appeal process can be viewed through the lens of frame theory, as introduced in chapter 5 (Snow et al. 1986). Similar to the way in which organizations frame issues to appeal to new members, coalition brokers seek to present issues to organizations in ways that will bring them into ideological alignment with one another. In the lexicon of frame theory, the task of brokers, at least in the early stages of coalition building, is frame amplification and extension. Frame amplification occurs as the broker attempts to heighten the importance of values and beliefs already held by the potential coalition partners in such a way that their concerns become more clearly aligned with those of others. For example, in recruiting unionists to work together with the environmental community, a broker might emphasize existing union concerns about health and safety, thus clarifying the connection between environmental hazards and those issues already addressed by worker advocates.

Frame extension takes place when an SMO tries to "extend the boundaries of its primary framework so as to encompass interests or points of view that are incidental to its primary objective but of considerable salience to potential adherents" (Snow et al. 1986, 472). A coalition broker might attempt to draw workers into alliance with environmentalists by citing the fact that environmental protection is often associated with job creation. Job creation is obviously not the core goal of environmental advocacy, but to the extent that it does serve the interests of workers in an incidental way, this kind of appeal might reasonably be used to win labor support for environmental measures. In an example presented in chapter 5, an Audubon Society organizer working on an antisprawl campaign did outreach to urban community leaders, citing the fact that city centers lose jobs and tax revenues when businesses relocate to the suburbs or more rural outskirts. Saving urban jobs was not on the Audubon agenda, but by extending the issue in this way, environmentalists hoped to appeal to other potential supporters of its antisprawl campaign.

As discussed in the previous chapter, more significant transformations may occur in the thinking of actors from both sectors that participate in a coalition once coalition work develops. Frame theorists refer to these more radical conversions as "frame transformation." Examples cited in the last chapter wherein labor leaders, who previously had given little consideration to environmental issues, came to see value in the preservation of nature for its own sake, represent frame transformation. Likewise, environmental advocates' taking a stand on the minimum wage or the right of workers to organize also represents an alignment of organizational frames that would be considered of the transformative type.

But this more significant type of transformation of organizational mindset does not typically occur in the early stages of coalition building.

Transformations of this type are best considered under the rubric of organizational learning. Learning suggests a deeper cognitive transformation compared to the relatively superficial appeal process that frame theory takes as its object of study. Brokers are framing issues for potential coalition partners in the same way movement organizations reach out to potential members. More fundamental transformations in an organization's ideology typically take place over a longer period of time, once movement actors have been routinely exposed to the ideas of their coalition partners.

One barrier in the frame alignment process between unions and environmental SMOs may be found in the differences in the broader conceptions regarding politics and social change held by the actors in these distinct sectors and a resulting shortage of those capable of effectively fulfilling the broker role. Whereas collective-action frames presented by particular SMOs have primarily been used as movement-specific interpretive orientations, "master frames" refer to broader conceptions regarding social problems, their causes, and solutions. These more generic orientations can be adopted by many different movements and have been offered as an explanation for waves of protest that occur simultaneously around a wide range of distinct issues (Snow and Benford 1992). There is some indication that, beyond the distinct issues addressed by the labor and environmental movements, actors within these movements have fundamentally different views of social problems and the nature of needed social change.

Having a common master frame can increase the probability of coalition activity between two organizations in two ways. First, it provides distinct movement organizations with a similar framework for approaching the issues they address. Even when organizations address dissimilar issues, if they share a common framework regarding the underlying source of social problems, they are more likely to find justification for mutual effort. Second, there are likely to be more movement participants who traverse organizational lines given the similarity of their political vision. It is these individuals who can play the role of coalition broker. Carroll and Ratner (1996), who have analyzed the association between master frames, movement organizations, and their participants, tested this proposition. They found that environmental-movement participants more often tend to embrace a "liberal frame" in which "the state is envisaged as the *container* of politics and . . . governments as powerful agents that *adjudicate* conflict among groups in society" (609). This is in contrast to labor organizations, in which participants rely more upon a "political-economy frame" in which power is viewed as structural and materially grounded.

Carroll and Ratner also found that those adopting a political-economy frame tend to be more involved with cross-movement organizing, especially in regard to movement organizations in which participants utilize a common master frame. The central implication of Carroll and Ratner's findings is that environmentalists and labor activists have a larger gap to bridge in terms of the political understandings under which they operate. Not only must brokers between these two groups bring about common understandings in regard to particular issues that may be foreign to one set of actors or the other, but they must also navigate more fundamental differences in the political orientation of these actors. This, in turn, suggests that relatively few people are situated in such a way that they can effectively reach out to both sets of actors. Thus the framing task of labor-environmental brokers may be particularly burdensome.

In addition to identifying stakeholders with the potential to be coalition partners and motivating them to participate in a coalition by framing issues in a way that resonates with their preexisting concerns, coalition brokers must also provide a forum for interaction; an actual meeting between the potential partners must be arranged. Although this may be a simple task that does not require in-depth analysis, it is a necessary element in the overall process of coalition building. Many coalition theorists analyze the more abstract configuration of organizational interests around a particular issue but neglect the fact that the mundane processes of organizational interaction must be carried out if cooperative action between organizations is to transpire. Simple issues such as scheduling difficulties or the geographic distance separating two organizations can make the difference between heated conflict and close intermovement cooperation. It is the coalition broker who typically addresses these issues.

Having examined the various roles a coalition broker must play and the tasks associated with them, the next question to consider is, who plays the role of coalition broker? The strategic framing and mechanical tasks have been identified, but research has also indicated that some individuals are more suited to successfully carry out these duties than others. In *Coalitions across the Class Divide*, Fred Rose (2000) examines coalitions formed by labor unions with peace organizations and environmentalists. Emphasizing the cultural barriers that exist between middle-class peace activists and working-class unionists, he identifies the characteristics of the interclass "bridge builders" who successfully act as brokers between the two. According to his analysis, those who are most capable of playing the role of coalition broker are those who in some way straddle the class culture divide. He identifies several categories of people who fit this general description: middle-class New Left activists who joined unions in the 1960s as part of an effort to build a radical working-class movement, middle-aged people who grew up working class but later established themselves as middle-class and became active in middle class causes such as environmental protection, and older radicals schooled in the New Deal or prewar socialist era when progressive causes were less fragmented than they are today. According to Rose, it is these actors who are capable of bridging the cultural gap between the labor and environmental movements. They are able to play the crucial role of defining issues in ways that are not alienating to either middle-class or working-class members of potential coalition partners, an ability others often lack.

Although I question the emphasis placed on class culture as a barrier to intermovement cooperation (see chapter 8), it is clear that brokers are likely to be more successful if they are able to present issues in ways that merge the concerns of environmental and worker protection. Someone of a similar cultural orientation may be more capable of framing issues in effective ways and of identifying the subtleties of presenting ideas in a nonthreatening manner than someone of a different cultural background.

The Carroll and Ratner study is again useful in regard to this issue. Although Carroll and Ratner found fairly stark differences in the political understandings held by labor activists and some environmentalists, a disjuncture that would pose problems in terms of generating a common agenda, they also found that great variation exists among participants in environmental organizations regarding the master frame they tend to utilize. In other words, participants from some environmental SMOs may more easily bridge the gap with organized labor because they have understandings of social problems and their solution that are in common with those of labor union members. This is consistent with the findings presented in chapter 5 in terms of organizational range. Although framing was not a central object of inquiry in this analysis, it is likely the case that environmental organizations with a broader range are also associated with master frames more similar to those used by labor activists. To the extent that some environmental-movement organizations frame environmental concerns in relation to their social consequences, as opposed to conceiving of them as strictly wilderness issues, they are utilizing a collective-action frame similar to that of unions, and there is likely to be a common master frame associated with it.

Thus some environmental activists may be more suited than others to reach out to labor union members. But given the fact that the general political orientation of union activists and environmentalists tends to be fairly distinct, it is also possible that certain groups outside of the two sectors will need to serve as an organizational bridge between unions and environmental organizations. Although individuals from organizations within one movement sector or the other may be able to play this part, in some cases actors from a third sector may be needed to fulfill this role.

Analyses regarding culture and the master frames embraced by actors in different movements are informative, but at base what appears most essential for successful brokering is that a potential broker between two movements have a basic familiarity with the issues and concerns of both. If distinct organizations are pursuing the same goals, there is an obvious and easy link that can be cited in an appeal for mutual effort. But if the issue alignment between the two is not precise, the broker must be able to link the established concerns of the two organizations with the issue to be addressed. This was the key issue of focus in this study.

Although most coalition study has neglected issues regarding the actual process of coalition initiation, the research discussed above suggests three areas that need to be addressed: the individuals who play the role of coalition broker, third parties rooted in organizations that in some way span the gap between unions and environmental SMOs, and the specific issues that are framed as common to both sectors. The following sections will explore each of these three areas in relation to case study evidence regarding labor-environmental coalition formation.

Labor-Environmental Coalition Brokers

Considering first the characteristics of the individuals who played the role of coalition broker in the cases examined in this chapter, each expressed concern for both union and environmental issues. Although some of the other leaders interviewed spoke only of instrumental gains to be had through cooperative work with organizations in the other sector or were dismissive of their concerns, those who played an active role in forming intermovement coalitions expressed strong interest in the issues of both sides. Of course, organizational learning can take place as a *result* of coalition activity, raising questions about whether the dual interest expressed by brokers emerged before or after their coalition involvement. But in most instances the personal background of the coalition brokers made clear their preexisting interest in both labor and environmental issues.

Thirteen of the individuals interviewed were primary actors in the initiation of intermovement cooperation and the formation of labor-environmental coalitions. Although several others reported making some effort to reach out across movements, their appeals either went unheeded or resulted in only minimal short-term cooperation. Of the thirteen I have identified here as coalition brokers, two were environmentalists, nine were union leaders, and two were organizers for a more general public-interest organization. The disparity in terms of the number of unionists serving as coalition brokers relative to environmentalists is consistent with the idea that a broader organizational range corresponds with more coalition activity. Unions, especially union federations, address a wider range of issues than environmental organizations; thus they are generally more able to initiate coalition work. The data available here cannot, however, be used to draw any firm conclusion regarding the propensity for coalition brokers to be found in unions as opposed to environmental organizations. Two of the spurned outreach efforts that did not qualify the person who made them as a practicing coalition broker here were made by environmentalists to labor leaders; if their efforts are taken into consideration, the margin of difference in the number of labor, as opposed to environmental, coalition brokers shrinks. Regardless, the cases examined here do not represent a random sample, nor is the number of cases examined large enough to support statistical generalization. The main observation to be made regarding the research findings presented in this chapter is that coalition brokers were found in both the labor and environmental sectors as well as in third-party organizations.

The dual concern for worker and environmental issues exhibited by several of the coalition brokers was evidenced not only in their rhetoric, but also in their organizational affiliations. Four of the union leaders who acted as brokers were involved in environmental politics independent of their union activity. Tom Fagan, former president of the International Union of Electrical Workers local in New Jersey and now a staff member of the international, had been involved with Greenpeace and currently sits on the board of directors of Clean Ocean Action. Gerry Gunderson, a union activist with the United Steelworkers in Wisconsin, is a member of the Wolf Watershed Education Project, an environmental group active in river protection issues. Ron Judd, the executive secretary treasurer of the King County Labor Council in Seattle, worked as an outdoor adventure guide and environmental educator before entering the labor movement. Judd currently sits on the board of the Washington Wildlife and Recreation Coalition, in what one environmentalist described as a labor-environmental "interlocking directorate." Jane Perkins served as president of Friends of the Earth (FoE), a national environmental organization, before taking a position with the national AFL-CIO as environmental liaison. Prior to her work at FoE, Perkins was a union activist with SEIU and a victim of the Three Mile Island nuclear accident. A fifth broker from the labor side, Ned McCann of the Maine Federation of Labor, has no direct environmental experience, although he considers himself an avid outdoorsman and is close friends with the chief lobbyist for the Natural Resources Council of Maine.

Two more coalition brokers from the labor community, Ed Ruff of the New York State AFL-CIO and Bob Duplessie of the fire fighters union in Maine were involved with health and safety issues at their unions.

They did not mention having any environmental background, but their work in the area of health and safety led them both to identify connections between the workplace and the external environment. As will be discussed later, health and safety issues serve as an important bridge issue for environmental and labor concerns. The last two labor brokers, Dominick D'Ambrosio of the Allied Industrial Workers in Wisconsin and Tom Quincy,[1] a staff member of a building-trades union in Maine, were essentially solicited to be brokers by other labor associates. Neither had extensive awareness of environmental issues prior to their experience as coalition coordinators and, not surprisingly, of the coalition brokers with union affiliations, these two placed the most emphasis on the amount they had learned from environmentalists through their coalition involvement.

One tentative observation regarding the five union brokers with environmental backgrounds is that they were generally younger than the other labor activists who played a broker role. Although this is certainly insufficient grounds for generalization, it suggests that further research on the cohort effect of labor leadership is in order. The new generation of union leaders now in their forties and fifties came of age during the turmoil of the 1960s and the rise of environmentalism during the 1970s, as opposed to the Cold War and the rapid growth in consumerism that characterized the formative years for older leaders. Some cultural theorists claim that New Left activism has shaped the approach adopted by young labor leaders in regard to environmental issues (Rose 2000). Survey research also supports the view that younger people are more likely than older people to have strong environmental sentiments (Jones and Dunlap 1992).

The two environmental leaders among the coalition brokers identified in the research presented here, Anne Rabe of the Citizens' Environmental Coalition and Amy Goldsmith of the New Jersey Environmental Federation, did not report having any direct experience in the labor movement. Their classification here as coalition brokers derives from the fact that they both work for environmental organizations whose range encompasses toxics issues as well as those that involve community health. When "community" is viewed as including workers whose health is put at risk by the same substances that threaten the surrounding environment of the community, the logic of interorganizational outreach by environmental organizations with a community health focus becomes obvious. The other two brokers identified here are situated in organizations that themselves bridge the labor-environmental divide. Their cases will be discussed separately in the context of looking at such bridge organizations.

The key finding derived from the majority of cases examined regarding coalition brokers is their prior concern for both labor and environmental issues. In almost every case the broker in some way straddles the divide between labor and environmentalism. This indicates that a simple overlap in issues of concern may not be enough, in itself, to generate cooperation between movement sectors, but that at least some modicum of preexisting sympathy in one organization or its leadership for the cause embraced by an organization in another sector is part of what motivates interorganizational outreach. Labor leaders with environmental sensibilities and environmentalists who are concerned with worker health are more likely to engage in outreach than those who simply find themselves working on an issue that also happens to be on the agenda of another organization.

Although formation of alliances as a result of a broker's outreach is the general pattern identified among nearly all of the cases examined in the research presented here, it is possible that intermovement bonds can be formed even in the absence of coalition brokers, though, such occurrences are rare. In fact, only one such case was identified among all of those investigated here, and it came about only in the face of a threat of very serious confrontation. Although this case is exceptional, it is worth reviewing to demonstrate the possibility for similar outcomes to emerge from diverse conditions.

The case involved New York Audubon, a land use organization that is not well integrated into the existing coalition of unions and environmentalists in the state. In the early 1990s Audubon launched its Listen to the Sound program, designed to build support for reducing the amount of nitrogen escaping into the Long Island Sound. Excess nitrogen in the water, in part caused by run off from development projects, can result in oxygen depletion, which threatens marine life. Audubon held a number of public hearings on the issue and was beginning to push for the implementation of nitrogen reduction strategies for the sound. It was at one of their public meetings that Audubon leaders became aware of the potential conflict that its push for the adoption of such measures might generate with organized labor. According to the state executive director for Audubon, David Miller:

The construction industry [workers] came to our second annual meeting out of fear that nitrogen controls would cause a building moratorium that would shut down all housing development in Westchester at a time when the construction industry was facing a 35 percent unemployment rate in the New York metropolitan area. They came on strong with 1,200 Teamsters picketing outside of our conferences in seven-degree weather. But we were able to begin a dialog with them the day before, so when the issue came forward, both of us were able to extend an olive branch. I guess the best analogy is two trains on a track about to hit each other. Both leaders of the trains were smart enough to stop them long enough to get out on the tracks and talk before they crashed. (personal communication, May 12, 1999)

Following the picketed meeting, Miller and the head of the building-trades unions entered into a dialogue through which they were able to arrive at a mutually agreeable solution. To improve the water quality of the sound, they agreed to pursue state and federal funds collectively for the upgrade of water treatment facilities and the restoration of wetlands, work that in both cases would be carried out by building-trades workers. In this successful effort at compromise cooperation, Audubon was able to pursue its pollution reduction agenda along with building-trades unions interested in the jobs that implementation of pollution reduction strategies would create. The organizations formalized this effort in the form of the Clean Water Jobs Coalition, of which Audubon is the only major environmental partner. This was an unusual development, given the lack of any obvious coalition broker facilitating formation of the alliance. Miller's willingness to redirect his organization's energies into raising funds for the water treatment facilities upgrade project and the general pragmatism expressed on both sides may have made this kind of agreement possible.

Thus, the core finding regarding the role of brokers who hold dual sympathies in the formation of labor-environmental alliances does not preclude the possibility of instrumental or even compromise cooperation among partners that otherwise have no preexisting sympathy with one another's goals. Although relatively rare, long-standing coalitions can develop in the absence of these important actors. This suggests that the cliché that "politics makes strange bedfellows" is sometimes valid. Yet in most cases the bedfellows found in political action are not strange at all. Usually alliances form among organizations with very similar concerns or at least among those whose goals do not normally conflict with one another. But on occasion, a particular policy may unite organizations and activists who otherwise harbor hostility toward one another's goals. The presence of an anomalous individual in an organization who happens to embrace both movements would make that organization more likely to engage in coalition work involving organizations from the other movement, as that individual can then serve as the broker that brings the two organizations together. But on rare occasions, even long-time adversaries may find the means to unite on a common issue without the aid of a mutually sympathetic broker.

A second caveat regarding coalition brokers should be added. The fact that most coalition brokers already hold sentiments regarding the issues important to the other side does not mean that organizational learning is unimportant in these instances. The case studies examined in the research presented here clearly demonstrated that expansion of organizational range does occur, even in the organizations these brokers lead, in part because active coalition participants themselves develop a still greater understanding of one another's concerns and in part because they are able to educate others within their sector about the importance of incorporating those concerns into their work. Organizational learning is probably more significant for individuals who are drawn into coalitions by brokers simply because they have more to learn, but coalition participation also had an educative effect on the brokers themselves in several instances examined here.

Framing Labor-Environmental Bridge Issues

As established in the previous section, coalition brokers tend to be those who hold sympathies with both of the movements being brought into cooperative relations. The next question to consider is, what issues allow for initial collaboration between coalition organizations? In most of the case studies examined here, early contact by the coalition broker involved an issue that would allow for instrumental cooperation between two organizations. Given that SMOs are generally focused on specific issue areas, it makes sense that the issues that first bring organizations from two sectors together are those involving policies that in some way address their distinct goals. One of the primary tasks of a coalition broker is to frame issues in a way that appeals to all of the coalition's potential partners, and starting with issues in which dissimilar organizations have an established identifiable stake makes the framing process relatively easy.

Coalitions were found, to be involved in issues ranging from watershed restoration to corporate-welfare reform to whistle-blower protection. But the issues that most commonly generated the original contact between unions and environmentalists were health and safety issues involving toxic substances. Labor and environmental organizations engaged in a wave of cooperative efforts on toxics issues in the 1980s, providing the first major joint labor-environmental effort for many of the groups examined. New York offers a typical example of the bridging process around toxics issues. Ed Ruff was cochair of and a key coalition broker in the New York Labor and Environment Network. He held the position of health and safety director at the New York State AFL-CIO at the time that he and Anne Rabe of the Citizens' Environmental Coalition, along with some local coalition groups, organized the statewide network. It was work on toxic contamination both in the workplace and in the surrounding community that first brought Ruff and Rabe into contact with one another. They quickly recognized the similarity of their concerns and the strategic benefit of coordinated action between their two organizations, and each worked to bring their organization into alignment on toxics and other issues.

As noted above, a key task of coalition brokers is to frame relevant issues in ways that appeal both to their own members and to their prospective coalition partners. Ruff and Rabe were largely in agreement regarding the problems associated with toxic substances. Yet to facilitate the formation of a statewide network to address those problems politically, they had to serve as translators who reinterpreted the relevant issues for consumption by different constituencies. Ruff describes his interpretation of environmental issues from the perspective of a worker advocate and his role as the translator of environmental concerns within his labor federation:

If you are working in a plant, with chemicals, and you are putting those chemicals, into the air . . . you may be fighting for safety and health laws to help the workers on the job, but who is protecting those same workers in the community? . . . When [workers] have a chance to work, they will not worry about the environmental stuff. They will worry about falling off a ladder or getting hit with a wrench. But they never think about what happens to the environment. . . . [For me] it wasn't just "I'll take care of the inside of this plant, I don't care what happens outside." You realize that you are only protecting the worker. Who is going to protect his family if you don't take up and finish the fight? [At the AFL-CIO] any safety and health bill or environmental bill that ever came in for our support would be sent to my office. I would review those bills and give my opinion to the legislative director, who would then write either in support or against those bills, based on my opinion as a safety and health expert. So when I got strictly environmental issues, the first person I would call was Anne and ask for her opinions and reasons why we should support those bills. And we became very close in supporting most of the environmental bills after that. (personal communication, May 12, 1999)

Here Ruff describes his role both as a liaison to the environmental community through his consultations with Rabe and his role as a kind of translator, taking "strictly environmental issues" and framing them as union safety and health concerns.

The monitoring and control of toxic substances played a central role in the development of cooperative relations in New York, Maine, and Wisconsin and survey data indicate that this is true of several other states, as well. Brokers and other coalition participants described toxic-substances control as the issue on which both sides could most readily recognize their commonality. With issues that involve a clearly instrumental benefit, the work of the coalition broker is relatively easy. The translating role that brokers must play between their own membership and their coalition partners is not complicated when interests align behind support for a single policy or if only minor compromises are needed to secure the goals of both sides. Ruff conveys with little difficulty why the union should support environmental measures. He amplifies traditional worker health concerns and in doing so focuses on the obvious link between workplace and community health. Brokers may be aware of and utilize certain subtleties when talking to different audiences in order to account for different political orientations or cultural backgrounds, but their fundamental task is to frame the issues in a way that has appeal for both sides.

The real challenge for coalition brokers comes when labor and environmental positions on a particular issue do not necessarily yield clear mutual benefit. Nonetheless, adept brokers are sometimes capable of extending frames in ways that make tangential or even threatening issues relevant to their audience, as Judd did in the example cited at the beginning of the chapter. He was able to frame an environmental issue designed to control development in a way that made sense to union workers in the building trades. Growth management, which had been viewed by these workers as a threat to construction industry employment, was presented instead as a means of rationalizing development and sustaining jobs in the long term. This view of growth management probably would not have been well received by those workers had it come from an activist in the environmental community, nor would an environmentalist be able to frame the issue in a way that had such appeal. This highlights the important framing role that brokers carry out, especially when they are promoting issues that do not readily generate mutual support.

A similar example of this can be found in Wisconsin, where Gerry Gunderson, a long-time activist with the United Steelworkers, mobilized the labor community to oppose an environmentally hazardous mining project. Gunderson was involved with the Wolf Watershed Education Project, an environmental group fighting to protect the Wolf River. A proposed copper- and zinc-mining operation, located near the headwaters of the Wolf River, threatened to contaminate the river with acidic waste. Environmentalists had mobilized against the mining project, yet Gunderson did not link his environmental sentiments with his labor activism until the president of a Steelworkers local in Milwaukee appeared in a television commercial sponsored by the mining company. The commercial touted the jobs that would be created as a result of the mining operation. According to Gunderson, "The catalyst for me was there being a union person in a kind of official capacity appearing in that commercial. He had the emblem of the United Steelworkers of America. I said to myself, well something's got to be done. This is not the official position of either the Steelworkers or labor and it certainly looked like that [in the commercial]" (personal communication, July 8, 1999). Working with other unionists involved in the Wolf Watershed Education Project, Gunderson formed the Committee of Labor Against Sulfide Pollution (CLASP) and rallied the rest of the labor community to take a stand in opposition to the proposed mine.

In his effort to promote labor opposition to the mine, Gunderson focused not only on the environmental threat posed by the project, but also on the poor labor record of the mining company and the negative long-term job impacts of environmental degradation. "In the local communities where mines are proposed," Gunderson notes, "they lose out on all the permanent tourism jobs that could be there, as opposed to the boom-and-bust nature of the mining industry" (personal communication, July 8, 1999). Gunderson was thus able to garner labor support by framing the grounds for opposition in ways more consistent with traditional union concerns, rather than focusing exclusively on the environmental dimension. He used his labor connections to target sympathetic unions, many of which signed on to the antimine campaign. In response, Gunderson says, industry stepped up efforts to demonize the environmental opposition and use job fear to win worker support:

The company likes the workers to think that [job loss in the mining industry] is because of environmental restrictions on mining. The company inundates the workers with propaganda. They even turned some of them out for a hearing on the mining issue. They gave them T-shirts, paid time off and even transportation to and from the hearing. It was held in the heart of Milwaukee's manufacturing district. They turned out substantial numbers, but not as many as the other side. A broad cross-section of union people turned out to speak in favor of the [antimine] measure. It was a major defeat for the company. They tried to dredge up as many people as they could, but there were still more in support [of the measure]. (personal communication, July 8, 1999)

The environmental background of coalition brokers from the union sector, such as that possessed by Judd and Gunderson, enables them to reach out to environmentalists regarding shared concerns, while at the same time serving as credible and convincing spokespeople when addressing their own membership on issues that do not typically fall within the range of union action. Although some issues more readily lend themselves to mutual support, brokers are often able to frame issues in a way that is consistent with the ingrained orientations of potential coalition partners. Yet in some cases organizational conditions inhibit the success of coalition brokers. A case in New Jersey offers an example of failure to achieve stronger labor-environmental ties even in the face of broker efforts.

New Jersey has been noted in earlier chapters as a state with poor intermovement relations because of the narrow range adopted by many of the major labor and environmental organizations and their lack of ongoing ties with one another, which might, if they existed, facilitate a broadening of organizational range. Efforts by some environmental groups such as New Jersey Audubon and the New Jersey Environmental Federation to reach out to labor have been thwarted by the narrowly focused state AFL-CIO. But even in cases in which labor brokers have tried to negotiate union support for environmental issues, little progress has been made on most issues, as in the case of labor-environmental conflict over toxic-waste disposal.

In the mid-1990s, Clean Ocean Action, a New Jersey–based environmental organization with a fairly limited range focusing on ocean pollution, found itself in a heated dispute with the longshoremen's union over the dredging of the major ports in the New Jersey region and the proper disposal of the accumulated toxic sludge that would be drawn from the ocean floor. Like the growth management issue in Washington in which Judd brought reconciliation, the New Jersey dredging issue was potentially one of jobs versus the environment. Periodically ports need to be dredged to ensure that they remain deep enough not to threaten the ships that move in and out of them. But in the New Jersey case, the material to be dredged was not just the silt and other organic matter that is removed as part of a typical dredging operation. Because of years of industrial activity in the area, the material that had accumulated on the ocean floor was laced with toxic substances. Without dredging, port activity would likely decline, along with the jobs it provided. But dredging meant that toxic sludge extracted from the ocean had to be disposed of, and environmentalists opposed the proposed plans simply to dump the waste out at sea. As a result, dredging advocates, including workers who feared job loss if dredging was not carried out, came into conflict with Clean Ocean Action and others who opposed the dredging plan.

Tom Fagan, a union organizer with the International Union of Electrical Workers and former Greenpeace activist, sat on the board of Clean Ocean Action. He had been personally involved in many environmental issues, and he had also moved his union to adopt proenvironment positions. As one of a few coalition brokers in the state, he attempted to mend the rift over the dredging issue:

> I think with the dredging issue, the union guys that I talked to up there felt slighted, because the environmentalists had put forth proposals or positions without trying to talk to them and finding out what their concerns were first. So I think both sides kind of got locked into their issues and didn't see the other side as well. That's why someone who's in labor and understands those issues all too well, but is also an environmentalist, [can play an important role]. I can kind of see both sides. I can definitely talk to [the unions] easier than Clean Ocean Action. At that point in time, they wouldn't have even talked to Clean Ocean Action. But as a labor rep, I can relate to what their concerns are better, so I put a call in to them and said, "Look, I hear what you're saying and I can help address those concerns and make sure those concerns are heard." (personal communication, May 27, 1999)

Although Fagan reportedly received a warmer reception than environmentalists had, leaders in the longshoremen's union were nonetheless resistant to his pleas for them to work with the environmental community, and Fagan's role was not decisive in this case. As Fagan suggests, both sides were already "locked in to" their positions. This is typical where there is little or no ongoing contact between unions and environmentalists. Compromises and mutual understanding between these two groups are unlikely to be achieved in the absence of their routine exposure to one another. When potentially divisive issues unexpectedly arise, neither is readily capable of conceding or even reaching out to the other because a pattern of contact and interaction has not been established. In this case, state lawmakers later stepped in and convened a working group that included industry and union representatives, environmentalists, elected officials, and other experts. The dispute was mediated through the working group, and it eventually arrived at a mutually acceptable solution.

Although the ultimate outcome in the New Jersey dredging dispute was acceptable to all parties, the case demonstrates some of the limitations placed on brokers when certain conditions have not been met. The narrow range of issues embraced by both the union and environmental organizations involved in the conflict limited the chances for an easy reconciliation of their concerns. Fagan was a more suitable representative of environmental concerns than environmental organizations when addressing union issues, but the lack of any ongoing contact between the relevant union and environmental organizations and the narrow focus that both have adopted created barriers that even an appropriate broker could not surmount.

Thus it is clear that coalition brokers can address certain issues more easily than others. Environmentally related health concerns serve as the most common initial bridge issue between labor and environmental organizations. Others can also unite workers and environmentalists, but skilled brokers must be present and other conditions must be met in order for more complicated issues to draw concerted action and to prevent disagreements between potential coalition partners from degenerating into open conflict.

Bridge Organizations

Given that toxics serve as the most common bridge issue between labor and environmental organizations, it is perhaps not surprising that brokers often hold positions in which they address these issues. For example, a union workplace safety official is likely to reach out to a toxics organization for help on health policy matters. This offers a more logical bridge than for that official to approach a land use organization about coalition participation (even though that same land use group may eventually be brought on board by fellow environmentalists in the coalition). This is consistent with the research discussed in chapter 5 regarding organizational range, which found that cooperation was most common among groups that had some amount of range overlap.

This also makes sense in terms of previous findings regarding master frames. Other research discussed earlier indicated that within some environmental organizations the "liberal frame" is predominant, whereas those within labor organizations tended to have a different understanding of politics. Although the study presented in this book focuses more on the issue areas of organizations, as opposed to their ideological predispositions, it is likely that as we move across the organizational-issue spectrum from groups having purely environmental concerns to those focused on issues of social justice, we would find a corresponding shift in master frames. Correspondence in terms of master frames has been associated with an overlap in personnel among organizations, thus we are more likely to find potential brokers among unions and toxics organizations than among unions and land use groups.

There are identifiable positions within each movement sector that a broker within that sector is likely to hold, for example a health and safety officer within a union or a toxics specialist within an environmental organization. In most cases the broker is someone holding such a position in one of the two primary sectors being brought into a coalition. In some instances, however, the broker role is filled by an individual from an organization that itself serves as a bridge between labor and environmental concerns. Committees on Occupational Safety and Health are quasi-independent labor organizations that focus on workplace safety issues. These are at times funded by labor unions, although in other cases they exist independently and receive funding from outside sources.

COSH organizations have been involved in labor-environmental coalitions in Wisconsin, New York, and Maine, and at times COSH leaders have played the role of coalition broker. In Maine, for example, Bob Duplessie of the firefighters union (currently a state legislator sitting on the Natural Resources Committee) was the vice chair of the local COSH, the Maine Labor Group on Health (MLGH), which played a broker role in efforts to gain passage of legislation at the state level restricting the use of toxic substances in the workplace. According to Duplessie, "We realize that toxics are going somewhere. So [our work] was a combination of workplace health at the work site for the employees, what they are being exposed to, and then what was going into the environment. So as the Labor Group goes, we were trying to look at the whole picture" (personal communication, May 20, 1999). This view of the "whole picture" typically leads to outreach and coalition building. The MLGH, and Duplessie in particular, played a significant role in bringing environmental and labor groups together in the fight for toxic-use reduction in Maine. Environmentalists were brought on board by what some may have viewed as a workplace safety issue. But because Duplessie could make a link between workplace safety and the external environment,

those groups lent their support. Duplessie's bigger challenge was addressing workers' concerns stemming from the job loss fear that employers
attempted to foster:

That was a real tough fight . . . the business community fought it so hard. [They
made claims] about how they were going to close and we were going to lose
thousands of jobs. . . . [When members raise the issue of job loss due to environmental protection] I ask them to think back when they were a little bit
younger. In the late sixties when we were at the forefront in terms of environmental laws, the paper mills in Maine said they were going to close. . . . Did they
close? No. They stayed in Maine. They complied with environmental laws. . . .
[Since then] they have created bigger mills, more production, but over the years
they kept cutting employees. We need to look at the full picture here. It's not
because of the environmental regulations that they're cutting employees, because
they're making more paper. They cut employees because of automation. . . . All
these mills are still operating and they've done good over the years. . . . It's not
the environmental regulations that are driving them out. (personal communication, May 20, 1999)

The Work Environment Council (WEC) in New Jersey serves as a
bridge between environmentalists and some of the more progressive
industrial unions in the state. Although the workplace environment
serves as its primary focus, the WEC has taken positions on some more
general environmental issues and has pushed unions to do likewise.

Progressive citizens' organizations can play a similar role in bridging
the gap between labor and environmental groups, and although they may
work on both environmental and labor-related issues, they differ from
the broker organizations just discussed in that they are not primarily
identified with one sector or the other. The Western States Center, based
in Portland, Oregon, is such an organization. Its mission is "promoting
and developing stronger leadership and greater capacity in citizens' organizations; creating new and lasting links between grassroots organizations and progressive public officials; developing a network of regional
leaders dedicated to solving critical public policy issues facing the West;
and advancing a Western agenda based on principles of social, economic
and environmental justice" (Mazza 1993, vi). The center was instrumental in uniting labor with environmental groups in opposition to the
wise use movement, an industry-backed effort to mobilize community
residents and rank-and-file workers against environmental regulation. Activists at the Center researched the funding trail of the allegedly

grassroots citizen and worker organizations that made up the wise use movement and were able to demonstrate to the union targets of the wise use appeals that conservative anti-union funders stood behind these front groups (Echeverria and Eby 1995).

Citizen Action, a network of citizens' groups, was mentioned by several labor leaders as being a bridge organization for labor unions to the environmental community. Labor and environmental leaders in Wisconsin both hold positions on the board of directors of Citizen Action and cite it as being a forum for labor-environmental meeting and dialogue.

In some cases these citizens' organizations, like the Western States Center consciously attempt to serve a uniting role among progressive organizations. They usually engage in a combination of work involving direct citizen empowerment and organizational coalition building. The Maine People's Alliance is an organization that employs such a two-pronged strategy, and the coalition-building portion of its work has contributed to developing interorganizational ties in Maine between unions and environmentalists.

In 1985 the MPA sought to "pick environmental issues that labor could be on board with to build coalitions . . . and to heal the divisions," according to one former MPA organizer. It strategically chose to organize around the issue of toxic chemicals, which it believed could unite labor and environmental interests. Leaders in the labor community who supported the MPA encouraged this endeavor, feeling that they themselves could not at that time establish direct ties with environmentalists. According to George Christy, the MPA's former director,

We understood that for any movement to really succeed, you need labor. So decisions were made where we could work as closely with labor as possible, build trust and solidarity, and join them in their fights. And in the meantime we had also been working with environmental groups to build this bridge . . . everybody is working to try to build bridges and build the trust in the state. None of this stuff happens by accident. We call it "spade work" here. We spend a lot of time preparing the ground. . . . We developed strong personal and working relationships [with both organized labor and the environmental community]. We worked together well for two years, so the ground was set where we could say, "OK, we trust each other, we don't agree on everything, but we can work together and we have a common interest, so let's do this thing." (personal communication, May 18, 1999)

Several labor and environmental leaders in Maine cited the mid-1980s antitoxics labor-environmental coalition as a watershed in the relations between unions and environmentalists in the state. The MLGH and the MPA, two third-party bridge organizations, were instrumental in building that coalition. The MPA has played a broker role in other instances as well. Environmentalists have at times sought out the MPA to make connections with labor when they felt they would not receive a warm reception as environmentalists. In another example, MPA representatives were designated to approach PACE workers at the Lincoln Pulp and Paper Company in an effort to seek environmental reforms within the company's plant. Such diplomatic missions can be a central part of the coalition broker's role, and bridge organizations like the MPA are better positioned to make the case for cooperation when latent tension exists between the two central movement sectors.

Industry-Labor Bridge Organizations

The role that bridge organizations play in facilitating interorganizational ties is not limited to social-movement groups. In fact, there is strong evidence that some industry leaders have recognized the value of creating bridge organizations of their own in order to strengthen ties between industry and labor around environmental issues. The wise use movement mentioned earlier can be seen as an attempt by industry to create organizational vehicles to serve that very function. Although the wise use movement includes a range of organizations that work on a variety of different issues, those organizations are united in their central message regarding individual private-property rights and opposition to federal regulation, particularly on environmental issues. Naturally, this has pitted them against many environmental organizations, whom they were in fact designed specifically to counter. The wise use movement started in the western United States, where industry and conservative foundations began to organize and sponsor groups that would speak out against environmental regulation. Since then it has branched out to create several national organizations with local chapters.

In addition to citizens' and property owners groups, one branch of the wise use movement includes worker organizations. These industry-funded groups attempt to frame environmental issues in classic job blackmail terms. Although they purport to support what they term a "balanced" approach to environmental protection, their positions tend to mirror those advanced by the industry itself. As discussed in chapter 4, one such organization was encountered in this study, the Pulp and Paperworkers Resource Council.

Started in the West during the spotted-owl controversy, the PPRC established branches in several paper mills in Maine in the mid-1990s. Although not officially linked with the union locals at which they have chapters, they typically approach workers through the union. PPRC organizers offer presentations to unionists about the dangers posed by environmental "extremists" and the PPRC's role in promoting "balanced" natural resource policies. They then seek union endorsement. If endorsed by union leaders mill owners then fund local chapters, which work as an appendage to the union. Although they identify the PPRC as a worker organization, PPRC representatives make it clear that the PPRC in no way supplants the union or coopts any of the union's work. Rather, they identify their efforts as supplementing union work in areas in which the union lacks the resources to address the relevant issues. They conduct lobbying and publicity campaigns around environmental issues supposedly on the basis of worker job interests.

The way the PPRC frames environmental issues clearly creates linkages between worker interests and opposition to what it refers to as "extreme" environmental measures. According to one PPRC spokesperson, "We consider ourselves environmentalists and we would like to see things done responsibly in the woods. But at the same time, we don't want to see certain laws get passed that are going to force businesses to close because they can't comply with certain time limits and things of that nature. . . . Before, our environmental goals . . . were starting to put people out of work or force companies to downsize or shut down completely."

PPRC representatives stress that they are not under the control of the companies that fund them. Yet local PPRC officials are given benefits not available to other mill workers, such as the opportunity to travel and do PPRC office work instead of working in the mill itself, and they are given paid time off to do lobbying or attend rallies. The PPRC also asserts that company officials take a hands-off position in regard to the internal functioning of the group and the policy positions it adopts. Yet despite this assertion and the group's close affiliation with union locals, its role as a link with management on environmental issues is evident. In explaining how it adopts positions on the relevant issues, a PPRC representative reported,

> We try to get the best information we can, though sometimes we do have to rely on what the industry is giving us for information. When they tell you that if this particular regulation goes through, they are going to have to cut down on operations . . . who is going to tell you that that's not the case? I would think that they would be in the best position to know how that's going to affect their industry. I don't think that they are feeding us bogus information. We give them the benefit of the doubt.

Another official added, "We have jobs that we need to preserve. And I guess it all boils down to who can be the best spokesman for the industry—a mill manager or labor? We can get into places they can't. That's why the system works. That's why the PPRC works."

Where the PPRC has been able to gain a foothold, it serves as an effective bridge organization between workers and management. Unions traditionally have an adversarial relationship with management, and workers often regard management assertions of job loss threats posed by environmental regulation with skepticism. The PPRC, on the other hand, is able to present itself as a worker organization defending worker interests. Citing its union ties, it presents what are essentially industry positions as being in the interests of workers, and many of those workers are receptive. According to one labor critic of the PPRC,

> The reason they say, "I'm with PPRC and I'm a strong unionist" is because it gives them credibility as representatives of workers. And they can't say, "By the way, the company is funding this whole thing and I get to fly all over the country for nothing," because it makes them look like maggots. They don't want to look that way, so they are very sensitive about portraying themselves as union folks. Officially and technically there is quite a distinction in my view.

During their presentations to workers, PPRC spokespersons effectively frame issues in ways that draw workers to the management side and away from environmentalists. According to the same critic, this has created some difficulties for union leaders seeking to build ties with the environmental community:

Some of our [environmental] efforts . . . slowed a little bit, because the local leadership was looking at rank-and-file [PPRC] activists working the other side. [They were] reluctant to hop into something that was going to be too controversial and toss them out of office. . . . It was a classic case of management co-opting the workers, and that has been a difficulty for us. I get nervous every time I hear of those folks, every time they show up someplace, about what the hell they are going to say.

Although some union leaders view it as a management front group, the PPRC has been successful at convincing at least some workers and union officials that it truly represents worker concerns on environmental issues. Its union ties offer credibility and the way in which it frames environmental issues draws workers toward the industry position. As noted in chapter 2, quantitative analysis demonstrates that a significant determinant of the quality of labor-environmental relations in a particular area is the quality of labor-management relations in that area. Cooperative labor-management efforts on environmental issues as manifested in the PPRC can sour relations between the labor and environmental communities. The situation in Maine offers a concrete example of the way in which this dynamic plays out and the crucial role played by this bridge organization.

Conclusion

The notion that organizations will instantaneously join together to address issues on which they share a common position does not correspond with the reality revealed by examination of actual cases. Nor is it valid to assume that coalitions will develop only when issues arise on which organizations have identical positions. In the first case, the notion is unfounded because it ignores many of the conditions that must be met for such coordinated action to occur. The second notion is flawed in that it ignores the possibility of conditions existing that can unite organizations around an issue even when those organizations' original positions on the issue do not perfectly coalesce. Among the central variables to consider here are the coalition brokers and the way they can actively bring about issue alignment and coalition activity.

Coalitions are common within movement sectors. Environmental organizations often work with other environmentalists, and unions, through formal relationships in the federation structure, commonly work together. The role of broker is relatively easy in most cases of intramovement cooperation, because of the natural alignment of different organizations on issues within that sector. Nonetheless the brokers' role in these cases is indispensable in making the necessary interorganizational contact and arranging for coordinated action. The true challenge in brokering, however, is for those who attempt to unite organizations from different movement sectors. The defining characteristic of most such brokers is their position straddling two distinct organizational perspectives on the same set of issues. In the cases examined here, that means an environmental activist with some labor ties or a unionist with an environmental background. On rare occasions organizations may manage to forge alliances with organizations from other sectors without the assistance of brokers, but in most cases, such actors play a key role in conducting outreach and framing issues both for their own members and for the members of the organizations they attempt to enlist as coalition partners.

Coalitions are likely to be built initally around work on issues that have obvious mutual appeal to the coalition's member organizations. The most common grounds for labor-environmental alignment are issues associated with workplace and environmental health. Toxic substances that threaten workers on the job also pose a threat to the surrounding environment. The elimination or the careful monitoring and handling of such substances represents a desirable goal for both union and environmental activists. But given the proper conditions, a skilled broker can inspire cooperation even on issues that have not previously been central to the missions of the organizations involved. Of course, certain conditions, such as when generally hostile relations exist or when organizations are very narrowly focused, can prevent coalition work despite the efforts of an adept broker.

Given that some specific issues are more amenable than others to concerted action, it is not surprising that actors holding positions in which they address those issues as part of their job duties more often play the role of coalition broker. Union officials who address health and safety issues or environmental leaders who focus on toxic substances are the most likely brokers for many labor-environmental coalitions. But sometimes, third parties play a role in bringing together activists from the labor and environmental sectors. General public-interest organizations can play a role in bridging the gap between these two sectors, especially when labor-environmental relations are tense and the two sides are not ready to deal directly with one another. This same gap-bridging role can also be played by organizations seeking to build worker-management ties. The Pulp and Paperworkers Resource Council serves as a bridge organization between unions and management, in some cases inhibiting the potential for a labor-environmental alliance.

In the final analysis, the study of coalitions must examine the key personnel who actually pull organizations into coalition activity. It is they who carry out the mundane, but necessary, task of interorganizational outreach. They also have a more difficult role in convincing actors in two or more organizations of their common concerns and the benefits of coordinated action among them. It is clear that certain characteristics make some brokers more capable than others at carrying out these tasks. According to the results of the analysis undertaken here, the necessary traits involve a knowledge of and concern for the issues held as important by those in the organizations that they hope to enroll as coalition partners. But these brokers face significant challenges as they attempt to facilitate intermovement cooperation.

Some scholars argue that a widespread convergence in the thinking of movement participants from the labor and environmental sectors is hampered by cultural differences between participants in these distinct movements. The next chapter takes up the question of whether a cultural divide exists between labor and environmental activists that affects their ability and potential to engage in cooperative action.

8
"That's Not the Way We Work": A Cultural Divide?

Tom Quincy (a pseudonym) worked for a building-trades union in the state of Maine. In the mid-1980s he was recruited by a retiring union associate to take his place on an advisory labor committee of a national environmental organization. In this capacity he served as a consultant to the environmental organization's leadership when it was considering campaigns on issues that would be sensitive for union members. This also brought him into contact with state level leaders in the environmental organization. Through his ties to both unions and the environmental community in the state, he was able to aid in several cooperative efforts on issues of interest to both unions and environmentalists. This often involved forest practices in the vast North Woods of Maine. Environmentalists were concerned about overcutting in the North Woods, and paper industry workers, whose union was among the most powerful in the state, had an interest in maintaining union jobs in the paper mills. Together they carried out campaigns against the export of unprocessed logs and wood chips, practices that both depleted the forests at a rapid rate and threatened union jobs at the local mills because of resource depletion and competition from foreign purchasers of these raw materials. But despite their ability to identify common interests and to work together on several campaigns, Quincy noted that differences in the political style of the unions and those of some segments of the environmental community occasionally created difficulties. He describes one instance in which he had helped to recruit members of his building-trades union to attend a public hearing and speak in opposition to a proposed wood chip facility on Sears Island, a campaign on which Quincy had been working with the environmental community:

It's a pretty interesting scenario when you get them into a room. . . . [T]here were just some really fine people from all walks of life. There is an association of people from islands along the coast, and they were all there. Those are mostly elderly people. And the fishermen were there. And everybody who thought that this might affect their lives. And the Earth First! people were there. . . . You had people dressed in expensive clothes and then you had Earth First! people dressed like hippies from the sixties. And some of them were dressed as salmon and eagles and one of them . . . was dressed up like the devil. They were heckling the governor and everybody at the hearing . . . And we had invited all of our membership to go to the hearing . . . so, a bunch of our members came. And some of our members, they didn't even want to [be involved with people engaged in these antics] given their background. You got guys who are fifty or sixty years old who aren't very tolerant of hippies. They didn't come up through the sixties like I did. So they were just grossed out [by the Earth First!ers] and they were ready to stomp out of the room. (personal communication, May 21, 1999)

Although Earth First! represents the radical fringe of the environmental movement and the building-trades union in question is among the most conservative in the labor movement, the general cultural clash between environmentalists and unionists as represented by this incident has received attention from those seeking to understand relations between these two sectors. Cultural differences in terms of political behavior can serve to divide movement actors even when common interests do exist.

Some scholars have linked these cultural differences to the distinct class backgrounds of the participants in the labor and environmental movements. As has already been discussed, participants in environmental SMOs come primarily from the professional middle class. Previous chapters have examined the way those in various occupational sectors are affected in different ways by environmental policies, in part explaining why some workers are more likely to oppose environmental regulation than others. Middle-class professionals are removed from the potentially harmful economic repercussions of most environmental policy, yet they still enjoy the benefits of a clean and healthy environment. Some professionals also benefit economically from environmental regulation. Thus a traditional class interest framework can at least in part explain disproportionate middle-class participation in the environmental movement and the potential for conflict between organizations made up of such middle-class members and working-class, blue-collar unions (Gouldner 1979; Murphy 1994).

Some have undertaken a more comprehensive look, however, at the way in which broad class differences may explain not only the differential economic impacts of state policy and the corresponding interests of different actors, but also the way in which political involvement itself, especially social-movement activity, has an identifiable class character (Eder 1993). One distinction, discussed previously, is the values-based politics of the middle class relative to the material-interest foundation of working-class mobilization. According to Ronald Inglehart (1990), the new values-based movements that attract middle-class participation are the result of the middle class's being of an economic status free from material want. The level of affluence achieved by the middle class enables it to turn its attention away from fulfilling basic economic needs and toward other values that find their expression in social-movement activity. Class differences in movement participation are therefore thought to reflect not just different interests, but differences in the cultural values and beliefs embraced by members of the middle and working class.

But the values-based politics of the middle class is considered different not only in content, but also in form, from the politics of the working class. New Social Movement theorists have elaborated on the distinct manner of expression and the organizational forms that new social movements take. Claus Offe (1987) argues that these new social movements differ from the "old politics" of economic growth and distribution, not just in their class makeup and nonmaterial basis, but also in their internal and external modes of action. These groups tend to conduct themselves internally in a way that is more egalitarian and informal than the hierarchical and formalized organizations representing traditional class interests. To use the extreme case, Earth First! boasts of not having an organization at all. Nor is there any formal hierarchy (or even membership). This contrasts sharply with the highly structured, hierarchical form of the labor movement.

Offe observes that new social movements express themselves externally through protest tactics and other unconventional approaches, in contrast the interest group bargaining traditionally empolyed in pluralist democracy. Again, the antics of the Earth First!ers described above are consistent with this characterization. Unlike Inglehart, who sees the expression of new values emerging in the absence of more immediate material needs, New Social Movement theorists offer a more structural explanation tied to capitalist development and the growing encroachments of technology and complex organizational systems into the "life world" (Eder 1993; Larana, Johnston, and Gusfield 1994; Melucci 1994). According to these theorists' explanation, middle-class activists create alternative organizations and at times turn to unconventional tactics to challenge a system that they see as inherently flawed and resistant to reform from within. This theory would explain not only the different issues being taken up by these activists, but also the unique organizational form and manner of expression of many new social movements.

But some challenge the macrostructural change explanation offered by the New Social Movement theorists. Fred Rose (1997) offers a critique of New Social Movement theory, suggesting that middle-class activism took a similar form in earlier periods, thus precluding late capitalist developments as a causal force in the emergence of new social movements. Instead, he grounds his explanation of political expression in terms of "class culture" generated ultimately by the different locations that the working and middle classes inhabit in the process of production. Since he specifically addresses relations between unions and middle-class SMOs, his assessment of the distinct middle- and working-class political cultures will be presented in detail in this chapter along with the work of other cultural theorists. This will be followed by an analysis of the value and limitations of the class culture explanation of inter-movement conflict based on case study material gathered in the research presented in this book. The critique offered here and the case study evidence gathered suggest that class culture plays only a small role in dividing middle- and working-class movements. Instead, I argue, attention is better focused on specific organizational features and the structural constraints placed on middle-class voluntary SMOs and labor unions.

Class Culture and Political Behavior

Borrowing from the work of cultural theorists such as Pierre Bourdieu (1990) and Paul Willis (1981), Rose (1997) argues that cultural distinctions between the middle and working classes are rooted in the organization of work. As he defines it, class culture consists of the "beliefs, attitudes and understandings, symbols, social practices and rituals throughout the life cycle that are characteristic of positions within the production process" (463). These cultural products of work life are then expressed in distinct forms of political action:

[S]ocial class shapes distinct cultural subsystems that order consciousness, organize perceptions, define priorities, and influence forms of behavior. The specific content of consciousness emerges through historical experiences and action within the framework created by class cultures. Movements reflect the class background of participants even if they do not explicitly articulate their goals in class terms. This has enormous implications for when and how people from different classes mobilize politically. (463)

As demonstrated by Harry Braverman in *Labor and Monopoly Capital* (1974), productive activity for the working class has in many ways been reduced to routine repetitive practices. Manual laborers have little autonomy or control over the production process, which is highly supervised by managers and regulated by the machinery the workers use. Direct supervision is accompanied by a system of rewards and punishments that creates, according to Rose (1997), "a [working class] culture based on compulsion. . . . Thus the culture of the work place is defined by the daily battle with authority" (476).

In contrast, middle-class professionals enjoy a great deal of autonomy in their work lives. Specialized knowledge is necessary to carry out the tasks with which they are charged, and direct supervision is inadequate to control this form of labor. Instead, professionals are in many ways self-motivated and seek to perform, not on the basis of immediate reward or punishment from supervisors, but rather for the fulfillment that they derive from performing their self-directed and more complex and mentally challenging duties. Thus, Rose (1997) says, "the middle class is organized around a culture of autonomy, personal responsibility, intellectual engagement, variability, and change" (477).

It is these differences in work life that create the foundation for distinct middle- and working-class cultures, which are then expressed in political action as well as other areas of life (Bronfenbrenner 1972). Manual laborers are in a constant struggle with supervisors who strive to extract more labor from them as workers attempt to resist. Because of their work experiences confronting authority, members of the working class come to understand political practice in terms of direct conflict of interests, and the political organizations they create reflect this understanding. Working-class organizations are formed to best unite common interests and confront political opponents motivated by interests different from those of these organizations' members.

Middle-class professionals, in contrast, apply their understandings of work life to politics as well, but for them political action is not a matter of conflicting interests, but instead a process of education and self-expression. According to Rose (1997),

Each culture produces characteristic beliefs about human motivations, politics, and social change as well as unique forms of organization and association. These differences are evident in the models of organizing advanced by each movement. In sum, working-class people live in a system of enforced authority, and they tend to approach social change through organizing around immediate, perceived interests. Professional middle-class life is regulated by internalized norms, ambitions, and responsibilities, and these movements tend to see change as a process of education about values. These differences produce two distinct class-based forms of politics and social movements. These different approaches to organizing are reflected in the issues that working- and middle-class movements pursue. Working-class labor and community-based movements generally focus on the immediate economic and social interests of members, while middle-class movements more often address universal goods that are non-economic. (478)

In summary, Rose argues that work life determines forms of political behavior. One's location in the class structure creates cultural predispositions that include the different ways class actors conceive of politics and undertake political action. For blue-collar workers, workplace antagonism yields confrontational interest-based politics, whereas the self-expressive nature of middle-class work gives rise to a politics based on education and intellectual development. Later in the chapter, I will examine the organizational differences between middle- and working-class SMOs identified by Rose and New Social Movement theorists, but first let us take a closer look at the class culture distinction and why, according to the class culture perspective, it generates intermovement conflict.

Class Culture and Timber Industry Conflicts

Examination of the class-cultural dimension is important not only for understanding differences between working- and middle-class social movements, but also for understanding why conflict can emerge between these two movement sectors. Rose examines a number of intermovement conflicts that pit middle-class activists against blue-collar unionists. Much of his work focuses on the military conversion movement and the unionized workers at weapons production facilities. Middle-class peace activists seeking to convert weapons production facilities to peacetime production met with great resistance from workers fearful of job loss. But Rose does not identify the basic threat to employment interests as the sole source of intermovement conflict in this case. Rather, the divergent class-based political cultures of the two movements led to misunderstanding and hostility.

In addition to the conversion movement, Rose also looks in detail at the timber conflicts of the Pacific Northwest in order to examine the class-cultural dimension of intermovement conflict. Because this case specifically involves unions and the environmental movement, I will use it in this discussion as a basis for examining the class culture issue. In many ways, logging encapsulates the cultural differences between workers and middle-class environmentalists, but as will be discussed later, it may also exaggerate them. Before addressing that, let us first consider the cultural distinctions thought to underlie intermovement conflict around this industry.

To begin, Rose argues that, like working-class laborers in general, loggers have an interest-based understanding of political action. The workplace antagonism between blue-collar workers and employers generates a framework in which loggers understand politics as competition between actors with identifiably different interests. This framework can lead to misunderstanding when loggers address issues that do not fall within the pattern of interest conflict they expect. Such is the case with environmentalists seeking to protect old-growth forests. As a result of their class-cultural framework, the values-driven concerns of middle-class environmentalists are incomprehensible to workers unfamiliar with a values-based approach to politics, setting the stage for misunderstanding and conflict. Workers used to fighting over concrete gain view with suspicion environmentalists' claims about an abstract concern for nature, with no clear interest base. As a result, loggers confronted by those striving to protect old-growth forests seek to identify the hidden interests that they believe must underlie the environmentalists' agenda.

Some may claim, for example, that wealthy environmentalists are scheming to protect the wilderness for their own vacation homes and recreational use, a suspicion sometimes confirmed by middle-class migration (Brown 1995). When no such interest assumption can be feasibly made, environmental motivations are dismissed as irrational or driven by spiritual fanaticism.

According to Rose's class culture analysis, the cultural divide between workers and environmentalists can drive workers to ally themselves with employers in environmental disputes. Despite the commonly conflictual relations between employers and employees, the actions of the employers, unlike the actions of environmentalists, are at least comprehensible to loggers within their cultural framework. Both sides openly strive to advance their own material interests. Employees' immediate adversaries, in the form of managers, are also people with whom workers interact on a regular basis and with whom they share many cultural attributes. Direct contact with management in the workplace and at the bargaining table also results in workers' having a level of familiarity with their employers relative to the total unfamiliarity of faceless environmentalists whose strategies often involve complex legal maneuvering from distant urban areas.

Cultural differences in regard to the basis of knowledge also serve as an arena of conflict for loggers and middle-class environmentalists. According to Rose, a central feature of logging, as of many working-class jobs, is that it is very task oriented, involving little in the way of abstraction or intellectual labor. This fosters an understanding among loggers that knowledge is based on experience, as opposed to abstract analysis, a perspective that is in direct conflict with that of members of the middle class, whose work experience places value on expertise and scientific study (Bell and Delaney 2001). The fundamentally different foundations upon which loggers and middle-class environmentalists base their understanding of forest issues makes mutual agreement on the source, or even the existence, of a problem difficult to achieve. Loggers believe, for example, that the forests are renewable, because of their experience clearing forests and then seeing them grow again. In contrast, middle-class environmentalists believe that the forests are in danger based on scientific reports generated by professionals like themselves.

Loggers also reject the claims of environmentalists regarding the unsustainable pace of forest cutting, because of the environmentalists' lack of direct experience with the forests.

Thomas Dunk (1994), in his cultural analysis of Canadian loggers in northwest Ontario, confirms this anti-intellectual tendency among timber industry workers: "The formal, abstract knowledge of the professional middle class is represented as being empty. Real knowledge is common sense and derived from direct experience" (28). Their fundamentally different foundations for understanding environmental conditions furthers the barriers to cooperation between middle-class environmental movements and working-class actors.

Rose identifies several other elements of logging culture that distinguishes it from its middle-class counterpart. He argues that the conceptions of nature held by the middle and working classes differ in important ways beyond their assessments of sustainable forestry practices. Loggers view nature as something to be used for human betterment and believe that in some ways logging practices actually improve on the forests' natural condition. Practices that alter natural systems and increase logging yields, for example, are seen simply as a better use of nature, and in no way an assault upon it. This contrasts with middle-class perspectives that tend to view nature as distinct from human activity and that regard any alteration of wilderness areas as an encroachment.

All of these differences, beliefs about human relations to nature, the basis of knowledge, and the appropriate form and content of political action, serve to drive a cultural wedge between loggers and the middle-class environmentalists who challenge logging practices. According to Rose, these cultural distinctions play an important role in the generation of labor-environmental conflict and the inability of these two movement sectors to resolve differences.

Like many others, Rose and Dunk assess labor-environmental relations by looking at conflicts within the timber industry, the most visible and highly contentious case of labor-environmental struggle. Such an assessment is useful for understanding those particular political conflicts; in many ways, however, it may distort our perceptions when we attempt to understand labor-environmental relations in general. National research on labor-environmental relations indicates that the tensions that exist between timber industry workers and environmentalists concerned with forest issue are particularly intense (Obach 2002). But using timber industry workers to examine the cultural differences embodied in social class may create a distorted image. Logging is in many ways not reflective of manual labor in general but instead represents a specific employment sector that has a number of differences from most blue-collar jobs. Several of these differences make logging and the corresponding logging culture much more prone to conflict with middle-class activists and thus a poor case to use as a basis for examining class-cultural differences in labor-environmental disputes. At least three important factors distinguish logging from most other blue-collar labor: (1) logging is based upon direct resource extraction, (2) logging work is contingent, and (3) extractive industries tend to be based in more rural, isolated regions. I will review each of these factors and demonstrate how they contribute to the distorted image of labor-environmental conflict and exaggerated sense of class-cultural differences presented by an examination of logging as exemplary of labor-environmental relations.

As noted above, Rose argues that blue-collar workers and the middle class understand the role of nature in fundamentally different ways. From the working-class perspective, nature is something to be used for human good. Trees are cut to make wood and paper and other products that meet human needs. Even the nature-related recreational activities more common to the working class, such as hunting and fishing, approach nature as a resource to be used by humans. Based upon this conceptualization of nature as existing for human needs, members of the working class develop the view that, with proper care and understanding of natural processes, human use of natural resources can be carried out indefinitely. In fact, in this view, human action can actually improve on nature in its implementation of practices that increase the yield that nature can offer. Human activity, from this perspective, is not viewed as a threat to nature. Although the evidence for the existence of this understanding of the natural world derives primarily from interviews with loggers, it is thought to be characteristic of the working class as a whole.

Such a conception of nature, to the extent that it can be generalized, however, may be more predominant among workers in extractive industries than among the working class generally. Logging involves direct contact with the natural world. Loggers may tend to view nature as a resource because, in their daily lives, they remove trees to have them transformed into usable products. The perceptions of workers in other industries, however, that involve less direct contact with nature are not likely to be as entrenched in the nature-as-resource perspective. Granted that workers in, say, the industrial sector are still working directly with natural resources, however, they have already been transformed in significant ways by the time most blue-collar workers encounter them. These workers do not participate directly in extracting these materials from nature, and their experience with these materials is in a form far removed from its natural state. If the basis of the class culture thesis, that work life fundamentally shapes one's belief system, including beliefs about nature, is correct, then one could hypothesize that workers in extractive industries possess an understanding of nature different from that of other blue-collar workers. If so, then those directly involved in the appropriation of nature for human use, such as loggers, serve as an unrepresentative sample from which to draw conclusions about the working class as a whole.

In considering the position of the middle class in regard to its view of nature and natural resources, professional work is still further removed from the direct use of nature than even that of most blue-collar workers. Most of this work involves abstract and intellectual processes as opposed to the manual manipulation of physical materials. The middle class more often experiences nature in a recreational context, in which the focus is more likely to be on aesthetic beauty than on nature as resource, a difference that facilitates the development of the different conception of nature that Rose and others identify. From the perspective of middle class professionals, the human use of resources is seen as an encroachment upon nature that should be minimized.

However, this distinction between the views of nature held by those in the working and middle classes may be overstated when we contrast professionals directly with extractive industry workers. These two groups are likely to hold the most disparate views since one group directly extracts resources from nature and the other is the furthest removed from the direct manipulation of material resources. Again, using the underlying basis of the class culture argument that work life shapes perceptions of nature, an alternative to a dichotomous view of the working class–middle class distinction in the conceptualization of nature is to think of perspectives on nature as existing on a continuum. The basic thesis about the way working- and middle-class individuals view nature may hold true for the extreme roles under consideration. Blue-collar extractive-industry workers may view nature as a resource to be used because of the character of their work life, and professionals may view nature as a pristine realm that should be preserved and utilized as little as possible, because their work activity is far removed from the manual manipulation of physical materials. But using this same framework, industrial workers are situated somewhere between extractive-industry workers and professionals on the continuum. They manipulate physical materials, but their work does not involve the direct appropriation of natural resources. If the class culture thesis is correct regarding the significance of work life for shaping ideologies, then industrial workers' views of nature should be distinct from both those of timber workers and those of middle-class professionals.

It may even be the case that workers in *nonrenewable* extractive industries, such as mining, hold views of nature that diverge from those embraced by workers in *renewable* extractive industries, such as logging, because of important differences in their work experiences in regard to nature. Miners see minerals extracted until they are depleted and the mine is closed. This may foster a different view of nature as a resource than that held by loggers, who level forests but then see the forests rejuvenate.

By extending the class culture argument still further, one might also hypothesize that working-class service-sector employees hold views even more similar to those of the professional middle class because they too are far removed from resource extraction and the use of raw materials. Food service workers or low-level office employees, for example, may be considered working class and may exhibit some cultural differences from professionals. But despite their different class status, there is no reason to suspect that the work life of these working-class employees entails a different relationship with the natural environment than that of their middle-class professional counterparts.

Thus the contrast in conceptions of nature between middle-class professionals and most members of the working class may not be as stark as that found between environmental activists and timber industry workers. The general class-cultural divide posited to exist in regard to conceptions of nature may be only a reflection of the unique position of timber industry workers. If in fact work experiences are fundamentally important in shaping the understandings of workers, as suggested by the class culture thesis, then we must take into consideration variation in workers' relationship to nature on the job. It is likely that, given their unique position as the direct appropriators of renewable natural resources, loggers conceive nature differently from most others in the working class. This is an empirical question that requires further examination. No direct evidence was gathered in this study to test these claims. The refinement of the class culture thesis suggested here is consistent, however, with the underlying principles of class culture theory itself in regard to the importance of work life for shaping cultural views. The central point in the refinement is that using loggers as a basis from which to generalize about working-class perspectives on nature makes class-cultural differences appear more extreme than they actually may be, thus exaggerating their role in generating labor-environmental conflict.

A second important feature of the logging industry that may exaggerate the probability of labor-environmental conflict when logging is used as a sample case is that logging is done on a contract basis, one job at a time, unlike an ongoing job in, say, a paper mill. In this respect logging is similar to work for many in the building trades, an employment sector often cited as another case for understanding labor-environmental conflict. Ron Judd, formerly the head of the Building and Construction Trades Council in Washington, explained the way contingent labor compels workers to focus on immediate concerns to the neglect of broader considerations such as the environment: "Unlike a factory job, where you are guaranteed work for the most part 365 days a year, year in and year out, in a building and construction trades job, you build something and then you have to go and build something else or else you're out of work. And so, it is all about building. Sometimes . . . we look at the short term too much and not the longer term sustainability of it, but it is about building" (personal communication, March 24, 1999). Just as the building trades are "all about building," the contingent nature of logging work makes it "all about logging." Economic insecurity is more acute for logging workers than for typical blue-collar laborers, altering the way loggers perceive threats to employment, including those presented by environmental advocates.

The very nature of contract work also creates a different relationship between employer and employee than that between most blue-collar workers and those who employ them. Contract work is inconsistent and dependent on one's relationship with an employer. Loggers obtain work in part based upon the reputation they have created for themselves. In a contract work system, workers and employers still have opposing interests on issues of pay and work, but workers have an incentive to minimize other differences with prospective employers. Even if a logger were to disagree with the practices of an employer or if they believed that cutting practices were not sustainable, they would be risking their livelihood if they were to side with environmentalists or to otherwise be viewed as troublemakers.

Again we can contrast this employment situation with that of most other blue-collar workers. For example, for those employed in the manufacturing sector, there is always the potential that a plant will be closed down and the workers laid off. On a day-to-day basis, however, workers enjoy relative job security. They have less of a need than contract workers to remain on good terms with their employer. Employees, especially unionized employees, who disagree with the practices of an employer can feel more secure in publicly voicing complaints about them. Of course, if a particular complaint is work related, employees may have recourse to a formal grievance procedure through which they enjoy legal protection against retaliation from employers. Even complaints about the employer's position on political issues not directly related to the workplace can generally be voiced, however, without the immediate fear of retribution. This is not necessarily the case for loggers and many building-trades workers. Given that the primary adversaries in environmental disputes are the industry owners and environmental advocates, workers who must remain in good standing with employers in order to remain employed are more likely to publicly side with those who control their access to work than are most workers, who enjoy a higher degree of job security. Again, the unique position of loggers in respect to continuing their employment sets them apart from most blue-collar workers and makes them a poor case to use for generalizing about working-class culture and differences with middle-class environmentalists.

A third major difference between timber industry employment and other blue-collar labor that makes timber industry workers problematic as representatives of all working-class members is the level of geographic isolation associated with timber communities. Rose acknowledges the isolation of timber communities and the suspicion of outsiders characteristic of such locales; this feature of timber industry employment needs to be given more consideration, however, as a confounding variable in the class culture analysis. Timber industry workers tend to live in small, isolated, rural communities that are often highly dependent on the timber industry for their survival and ongoing economic viability. Compared to blue-collar workers living in larger, less isolated, more economically diverse environments, geographically isolated blue-collar workers are likely to develop a cultural and sense of identity more distinct from that of the professional middle class (Brown 1995; Carroll and Lee 1990; Dunk 1994). Especially in communities that are greatly dependent on the timber industry, this isolation is also likely to generate a greater sense of threat from outsiders who seek to influence or restrict their economic activity in some way. Geography, then, is another confounding characteristic that complicates assessments of class culture in regard to labor-environmental disputes.

Dunk (1994) confirms the interaction of geography and class in this regard. In his analysis of logging culture, he argues that loggers' differences with middle-class environmentalists are not solely class based: "Environmentalism and environmentalists are connotatively linked to class and regional oppositions so that they become . . . symbols of other class and regional conflicts" (14). Forest workers, he argues, see environmentalists as "people from outside the region who do not understand local environmental or economic issues. Environmentalists are outsiders, and in particular they are people from 'down south' or from big cities" (24). Similar sentiments were found among geographically remote paperworkers interviewed in this study, as demonstrated by the following exchange, which took place during an interview with several paper industry workers affiliated with an antienvironmental paper industry organization in the state of Maine:

Bonnie [pseud.]:   The majority [of the environmentalists] are not from the area.

Ed [pseud.]:   I think all of them.

Norm: I think a lot of what's going on now as far as the direction that Maine is going is where a lot of environmental policy is being controlled from the southern part, because that's where the population is. Northern Maine is so sparsely populated. The people in Portland and the coastal areas are going to have a lot more say in what goes on in the state of Maine. And if they are not exposed to the life up here, they will probably listen more carefully to what somebody from out of state has to say about their proposals.

Bonnie: It's true. What's happening is we see a lot of Maine being bought up by out-of-staters, in particular in the coastal areas. Wages in Maine are not all that high. Most Maine people cannot afford to buy a home on the coast, and most of those homes are bought up by people from Massachusetts, New Jersey, Rhode Island . . .

Ed: . . . flatlanders . . .

Bonnie: . . . right, and then they start fueling fires.

Ed: That holds true at lakes and ponds all around this area. You go to any lake and pond around here, and I'll bet you a good chunk of homes there are owned by people out of state. The waterfront property is so expensive, it's tough to make ends meet here. (personal communication, May 19, 1999)

It is clear from this exchange that regional differences are a major feature of these paperworkers' opposition to environmental action. "Flatlanders" are not just people of a wealthier class, they are people "from away." Thus, it is not class alone that is responsible for the divisions between workers and environmentalists. Those in the timber industry and other geographically isolated blue-collar workers hold a chain of perceptions that place them in opposition to environmentalists: "southerner," "city dweller," "environmentalist."

Similar cultural barriers can be seen in other employment sectors, such as that of mine workers. Often they live in equally isolated, rural communities that are economically dependent on the mining industry. Perhaps not coincidentally, the United Mine Workers has been the most resistant to cooperation with environmentalists in recent negotiations between union and environmental leaders at the national level. The sense of geographic and occupational identity and opposition to outside influences is stronger among these workers than among workers who live in more economically diverse communities and who regularly interact with people outside of their particular industry. The stark cultural contrasts presented in class culture theory as general differences between the working and middle classes may not apply as readily to other working-class employment sectors as they do to the timber industry that Rose and others use as the basis for their analysis.

The opposition to middle-class environmentalists found among timber industry workers is overdetermined and not reducible to class culture. Their unique position among blue-collar laborers of being in an extractive industry, of relying on contingent work, and of being geographically isolated makes them a poor case for drawing conclusions about class culture generally. These characteristics make them more likely to have conflictual relations with middle-class environmentalists than virtually all other blue-collar workers. Thus, conclusions drawn from analyses of timber industry culture may be accurate for those communities, but they are not generalizable to the working class as a whole. This suggests that other factors besides class culture must be given greater consideration in the examination of labor-environmental relations.

Economic Change and the Declining Significance of Class Culture

The use of logging as a basis for understanding class culture and its relation to environmental conflict poses problems in itself, but there are other considerations that may render the focus on culture even less relevant to explaining the labor-environmental divide. First, in the U.S. context, to the extent that broad class-cultural differences do exist, they may be diminishing as the economy undergoes fundamental changes. The number of people employed in the manufacturing and resource extraction sectors is declining as a result of two factors. The first is that manufacturing work is increasingly shipped overseas to take advantage of the lower wages and less stringent regulatory requirements available in other countries. As the number of manufacturing jobs declines, a larger proportion of workers in the United States are employed in the service sector of the economy. The largest unions are no longer the industrial unions, such as the UAW or the United Steelworkers, but rather the Service Employees International Union, the United Food and Commercial Workers Union, and the American Federation of State, County and Municipal Employees.

Some professional jobs, such as teaching and nursing, are included in the service sector. There are many more types of employment in the service sector, however, that would not be considered professions, such as hotel employment or retail outlet work. Although many of these lower-status service jobs are routinized and highly supervised, like industrial employment, the nature of the work is qualitatively different from industrial employment. As already noted, workers in the service sector do not extract or even handle raw materials, making them more likely to view nature in terms similar to that of middle-class professionals than in terms similar to blue-collar workers. In addition, service-sector jobs are not task oriented like positions that require the assembly of parts or the felling of trees. Although the intellectual demands of blue-collar labor are often underestimated, there is likely to be somewhat more in the way of abstract or intellectual labor in low-level service positions where workers more often have to interact with and service clients or customers. Such workers are likely to have to contend with a wider range of circumstances than typical blue-collar workers and to assess how best to carry out their work. The task-oriented nature of traditional blue-collar work and the absence of job duties requiring abstraction are fundamental characteristics identified in class culture theory as cultural barriers between the working and middle classes. Even if this claim is valid, however, as service employment grows, the sharp distinction between working- and middle-class labor, and in turn class-cultural differences, will decline.

The second factor influencing the decline of manufacturing employment is automation. As manufacturing continues to become more automated, manufacturing jobs themselves are undergoing changes that alter traditional work practices and if the basis of class culture theory is correct, the corresponding cultural tendencies. As more jobs are automated, we see the amount of manufacturing employment decline; furthermore, those jobs that remain in some cases become more technical and complex. As the automation trend continues, fewer workers will be occupying positions on an assembly line carrying out routine mechanical tasks and more will be employed operating computers or servicing complex machinery. Even extractive industries like timber harvesting have been transformed by technology and automation (Bailey 1999). The use of this new technology is likely to generate jobs that call for more training and technical skill and generally more mental work than similar jobs used to require. Here again we see a reduction in the sharp distinctions previously drawn between working-class jobs and middle-class professions. Some of the claims made by theorists regarding class-cultural differences may be valid. Even if they are, however, they may also be of declining significance in terms of the role such differences play in dividing the working class from middle-class environmentalists.

Class Culture and Organizational Form and Practice

As established earlier in the chapter, timber industry employment offers a distorted and extreme case for contrasting middle- and working-class political culture. In addition, changes in employment patterns in the United States may render class-cultural distinctions less prominent. Yet despite their being overstated, some distinctions are apparent not just in the beliefs and understandings of middle- and working-class political actors, but also in the forms that political action will take among organizations representing the working and middle classes, an additional feature that can inhibit intermovement collaboration. New Social Movement and class culture theorists commonly cite the hierarchy and formality that characterizes old social movements relative to the egalitarianism and informality of new social movement organizations. Working-class movement organizations are thought to be task oriented and to engage in little abstract analysis reflecting the work life of those who make up the movement. Hierarchy and formality enables these organizations to advance basic constituent interests effectively and to "get the job done." Middle-class activism is thought to reflect the abstraction and personal expression that is central to work life. Participants seek fulfillment from joining political efforts by expressing personal values and beliefs. According to Rose (2002): "The forms of new social movement organizations also emerge directly from middle-class culture. Middle-class movements must be flexible and egalitarian to accommodate many individuals searching for their own identities and seeking a sense of purpose tied to their knowledge and actions. The emphasis on equality is an acknowledgement of the value placed on the individual quest to define one's own direction" (483–484).

The greater flexibility and more egalitarian nature of environmental organizations were confirmed by many of the subjects interviewed in this study. Several of the environmentalists interviewed noted the hierarchical nature of labor organizations and the difficulties that they encountered through their failure to recognize this difference between labor and environmental organizations. According to Anne Rabe of the Citizens' Environmental Coalition in New York,

The unions are very structured and very hierarchical. If you go to a rank-and-file person and ask them to go to an event, and it didn't go to the president, it could be a really big negative in terms of your future relationship with that union. . . . There are all kinds of appropriate channels you have to go through when approaching unions. We are nonprofit, grassroots-based. We have a certain amount of hierarchy in our membership and board, but it is much looser in how we communicate. (personal communication, May 11, 1999)

Amy Goldsmith of the New Jersey Environmental Federation reported a similar experience of unions' hierarchical structure in her efforts to organize union-environmental meetings: "There is a very clear hierarchy. It's very clear that if certain people aren't on board and with you, it's not going to fly. No one will show up. It needs the blessing of the president or the secretary-treasurer of that local . . . it has to come up from the hierarchy. While [in my organization] I am the hierarchy. I have to go to my board, but once I get [the approval of] that, I have a lot of latitude" (personal communication, May 26, 1999).

Ed Ruff of the New York State AFL-CIO also cited the lack of centralized structure and hierarchy among environmental organizations as a difficulty that he faced in his dealings with them. According to Ruff, "A lot of times they were speaking for individual groups instead of united. Each one had their own little piece, while the labor people spoke as a full body. We knew where we were going to be. I basically led, and the rest followed" (personal communication, May 12, 1999).

Subjects also confirmed that environmental organizations are more fluid in their issue concerns, offering support for the notion that environmental organizations are flexible and that they respond to the changing interests expressed by members. Claire Barnett, director of the Healthy Schools Network, an organization working on environmental-health hazards that affect schools, has solicited support for this cause from both unions and environmental organizations. According to Barnett,

There is clash of cultures between environmental organizations and unions functioning in New York state. . . . What I mean is the operating style of the organizations. . . . It takes unions longer to get on board an idea or a cause, but once they do, they stay with it. They are very loyal. Environmental organizations don't behave that way. They jump on issues immediately, they ride them for a while, they see where it goes . . . then they'll drop it and move on to something else. (personal communication, May 12, 1999)

The shifting focus in environmental organizations from issue to issue presents a problem for labor unions, which tend to have longer-term goals and expectations for ongoing commitments. George Christy is the chair of the Dirigo Alliance in Maine, which has brought together unions, environmental organizations, and other progressive groups to work on a number of issue campaigns and electoral politics. He commented on the tendency for priorities to shift within voluntary progressive organizations: "That's where labor really gets pissed off. 'One year you work with us and all of a sudden you leave. . . . We were supposed to work on this thing for three years, and now you guys don't want to do this any more.' That makes things difficult" (personal communication, May 18, 1999).

That priorities within middle-class organizations readily shift is consistent with the cultural explanation offered by Rose and the identity-seeking emphasis of New Social Movement theorists. The interests identified by working-class activists are less prone to change over time, whereas the expressive politics of middle-class activists are more subject to the whims of different participants.

There was additional support for class culture claims regarding the expressive nature of middle-class movement participation when subjects discussed the actual meeting processes. Union leaders identified the tendency for environmental representatives to stray from topic to topic or to seek to address larger questions about their vision of the movement, whereas union representatives sought to focus on a specific issue. According to Ruff, "The environmentalists sometimes followed the map of the issues, and we used to draw them back in and say, 'We know you might have an issue with the nuclear plant in, let's say, Norwich, but that's not the issue we are looking at today. We are looking to put a conference together that's going to address the Kodak issue, and we really need to focus on that.'"(personal communication, May 12, 1999). The task-oriented nature of working class politics does not mix easily with the expressive culture of middle-class activism. Christy also remarked on the clash of organizational styles at joint labor-environmental meetings:

Progressive organizations always have a couple of people that kind of have their heads in the clouds and talk about very high-minded kinds of things. And you look over at the labor guys and they'd be like "What the hell?" So that kind of stuff happens. The people in the labor community are drawn into very day-to-day things. They make things for a living. They provide services. They are not paid to sit around and have grand thoughts about how we are going to remake the world. These are the people who are building the roads and making the beds, so I think they get a little impatient at times with some of the folks in the progressive community. (personal communication, May, 18, 1999)

He further observed that conflict and majority rule procedures are common at labor meetings but rare among other groups he works with:

For the most part labor has a different way of operating than our progressive organizations. The labor movement is about conflict and about struggle. You go to their meetings and they have nothing in common, they have drag-out fights and they leave, put their arm around each other, buy each other a beer and they are friends. Most progressives don't fight that way. They want to have consensus. They are uncomfortable with face-to-face conflict. So when you bring those two organizational cultures together, they have a very difficult time relating to each other. (personal communication, May 18, 1999)

Again these organizational practices are consistent with cultural theorists' claims about class culture as reflective of work processes, with blue-collar workers focused on the task at hand and the interests they seek to promote, whereas middle-class environmentalists focus on expression and process.

Much of the case study evidence presented here supports the claims of class culture and New Social Movement theorists in terms of organizational form and practice. The hierarchy, task orientation, and interest focus of labor organizations contrasts with the egalitarian, expressive, value-centered nature of many middle-class environmental organizations. Furthermore, to some extent, subjects in this study did suggest that the differences identified posed difficulties in forging intermovement cooperation. Later in the chapter I will discuss the extent of those difficulties, which are generally minimal, but first I examine alternative explanations for these differences. Cultural explanations suggest that difficulties will arise in any interclass movement relationship. If the differences are rooted in something other than culture, however, then the broad claims of class culture theory about the barriers to interclass cooperation are unjustified.

Structural Constraints on Organizational Form and Practice: An Alternative to Class Culture

Despite the fact that some of the findings presented in the previous sector are consistent with the class culture perspective, further examination suggests that the contrasting practices and organizational forms of labor and environmental organizations may not be rooted in class culture but rather may be a reflection of the distinct positions of these organizations in the political structure. Legal requirements and the means available to these different types of organizations for obtaining resources and members necessitate distinct organizational forms and practices (McCarthy, Britt, and Wolfson 1991). In the following sections I will examine the distinctions previously identified as cultural in nature and offer alternative explanations rooted instead in the structural location of each type of organization. Such a structural explanation suggests that in some ways a more immutable barrier than class culture exists between at least some middle- and working-class organizations, preventing intermovement cooperation. Cultural differences can be overcome though learning on the part of the actors involved, but the elimination of structural barriers would require a more fundamental reordering of political representation and advocacy. I will argue, however, that at the organizational level of focus here, there is not a great distance between the structural locations of the labor and environmental sectors and that there is ample opportunity for cross-class coalition building. Structural barriers to organizational cooperation do exist, and various legal and organizational reforms would make coalition building easier, but a great deal of coalition work is still possible within the existing system.

Unions are commonly used as the basis for examining working-class political mobilization and in particular the interest-based orientation thought to characterize working-class movements in general. Yet the extent to which union structure or action can be explained in terms of cultural predispositions is questionable. Unions are regulated by the state and federal government and must adhere to certain requirements to remain in good standing (Summers 1977). The 1959 Labor Management Reporting and Disclosure Act includes a "Bill of Rights of Members of Labor Organizations" that includes, among other things, a requirement for unions to have a democratically established governing structure. Unions can be placed under government trusteeship if they fail to meet this requirement (Kannar 1993). Hierarchy and majority rule decision making in unions are therefore not simply a reflection of some cultural predisposition, as some have argued. Rather, these procedural and organizational forms are the product of legal requirements governing union activity.

Legally, nonprofit organizations such as environmental SMOs are considered "public-benefit organizations," as opposed to "mutual-benefit associations" such as labor unions (Bucholtz 1998). Public-benefit organizations face a different set of state regulations. Although public-benefit organizations do face certain constraints on their political activity with which they must comply to maintain their tax-exempt status, there are few legal requirements regarding how they are governed. To qualify as a nonprofit organization for tax purposes, organizations must create a board of directors to oversee their operation, but there are few requirements beyond that (Connors 1997). Nonprofit entities are run in a manner similar to private corporations, except that there are no shareholders to collect profit, and generally board oversight is even less stringent than that of for-profit corporations (Brody 1996). Given the different regulations under which they are governed relative to labor unions, the greater flexibility and informality of environmental SMOs is not the result of a middle-class culture, so much as it is of their freedom from legal requirements to operate otherwise. This is especially true of local grassroots environmental organizations, which in many cases are not even formally registered as nonprofit organizations with the Internal Revenue Service (McCarthy, Britt, and Wolfson 1991).

Union practices and goals are also shaped by their legal role in the representation of workers. " 'Mutual benefit' societies . . . are formed for the express purpose of advancing some interest, cause or goal shared by their discrete membership and not by the public generally" (Bucholtz 1998, 558–559). The primary way in which unions carry out their function as the representatives of their members' interests is through collective bargaining, another area that is directly regulated by the state. The state not only regulates the procedural aspects of collective bargaining and the tactics that unions can legally use to advance their interests but sets the parameters within which such bargaining can take place. The National Labor Relations Act specifies the mandatory subjects of collective bargaining as the "wages, hours and other terms and conditions of employment" (Mathis, Lawson, and Parker 1993). Unions are also legally required to represent workers' interests through an established grievance procedure when rights are perceived to have been violated. In short, unions are in many ways legally required to act as the representatives of worker *interests*. Thus the interest orientation theorized to reflect cultural predispositions is actually mandated by law.

The central point of the discussion here is that legal limitations on unions and their unique role as workplace representatives of worker interests must be given greater consideration when one is attempting to understand their operations. The interest focus alleged to be characteristic of working-class culture cannot be inferred from the structurally delineated organizational behaviors pointed out above. This is not to suggest that such an interest focus does not exist within the culture of the working class. Indeed, it has already been acknowledged that many unions in the United States tend to focus on servicing the workplace interests of members to an extent greater than that which is legally required. However, their propensity to focus on these interests is better understood as a result of legal and structural mechanisms that favor such a political approach than as a result of a cultural tendency characteristic of the working class.

The class culture explanation is even more difficult to maintain when we contrast U.S. unionism with that found in other nations. For example, blue-collar workers in Europe are engaged in similar production practices to those of their American counterparts. Class culture theory would suggest that this corresponding work life should give rise to similarly interest-focused unionism. Yet the "social unions" of Europe are often engaged in broad political struggles that bear little relation to the immediate interests of members. Thus the interest focus attributed to the political culture of the working class is better understood in terms of legal requirements and the political and organizational structures under which unions operate in different nations.

The class culture explanation is further undermined when we examine union activity at the state and national levels, as opposed to the local level, where collective bargaining and the representation of workplace interests are the primary union responsibilities. The federated labor organizations at the state and national levels do not face the same constraints as union locals do that limit their focus to workplace issues. State federations, the national AFL-CIO, and even many international unions engage in political activity that extends well beyond the interests of members and can in many cases be seen only as the values-driven pursuit of public goods. Political action on issues such as health care, welfare, education, and the environment, although always in some way tied to the well-being of working people, cannot be captured in the crude interest framework proposed for the working class.

There is even evidence to suggest that contrasting approaches to political action can be found among individuals who occupy different positions in the same organizational structure. As discussed previously, some theorists have suggested that those who hold higher positions in an organization's hierarchy are organizationally better situated than those lower down to make broader values-based decisions (Simon 1976). Leaders of union locals who are carrying out collective bargaining embrace an interest orientation because this is what is required of those in that position. Their focus is drawn to core interests, the wages, hours, and working conditions of their particular members, as is required in collective bargaining. But leaders higher in the union hierarchy are not constrained in the same way. Despite having a common class background with leaders of union locals, their position in the organization allows, and in some cases may even necessitate, a broader, more encompassing approach to political action.

Some of the subjects interviewed for this research confirmed that an interest orientation is more associated with union locals and a values orientation is more characteristic of encompassing federations. Several union leaders suggested that attention to parochial interests was more apparent at the local-union level, where the unions most often carry out collective bargaining. Dominick D'Ambrosio, president of the Allied Industrial Workers International Union and a long time labor-environmental coalition builder, commented on this tendency: "I think the higher up that you went the fewer disagreements that you had [with environmentalists]. I think this is probably true of most issues. The farther down the ladder you go in the organization you get [more narrowly focused agendas] . . . the people who were leaders had a much broader view of issues and it was surprising how many things we agreed on [with the environmentalists]" (personal communication, 1998). This view undermines the claim that a narrow interest focus is characteristic of the entire working class. Rather, organizational and structural factors tend to favor an interest orientation among those at lower levels of the union hierarchy, whereas those at higher levels, not facing such constraints, tend to adopt a broader values-based politics.

Aside from the broader values-based orientation of labor federation leaders, when we examine working-class political action outside of the union framework, values-based examples of mobilization can also be found. The antiabortion movement is one significant example of working-class political activism that does not fit within the interest paradigm.

In short, the labor movement serves as a poor basis for drawing broad conclusions about working-class political and organizational culture. Legal and other structural constraints limit the kinds of activities in which unions can engage at the local level. They are legally charged with protecting the workplace interests of their members through collective bargaining and grievance procedures. We see similar behavior among unions representing middle-class employment sectors such as teaching and nursing. These middle-class labor organizations, which represent an increasingly large proportion of unionized workers, are bound by the same rules and regulations as blue-collar unions, and not surprisingly, we see similar forms and behavior among these organizations despite the distinct class background of the membership. At the local level they seek to advance the core interests of their members through collective bargaining, and at higher levels of the union hierarchy, a wider range of values-based goals are pursued. The same pattern can be found among traditional blue-collar unions. Once we move beyond the local level with its circumscribed interest role, we see that union bodies behave on a broader, sometimes values-based framework. Although they are still largely working-class organizations even at these higher levels, we do not see the same limited interest orientation, because they do not operate under the same structural constraints as union locals. Thus it is the structural location and organizational features of labor unions that are primarily responsible for their political approach, not class cultural tendencies.

The other case commonly used to support the interest-based orientation of working-class political action is that of the environmental-justice movement (Bryant 1995; Bullard 1993; Gottlieb 1993). Environmental-justice organizations, populated by the members of the working class, tend to focus on immediate threats to health and safety presented by environmental degradation. They are often presented as a contrast to middle-class land use organizations, whose agendas are less tied to private-interest pursuits and represent more a values-based expression of environmental concern.

Organizing around direct environmental threats, however, is not unique to the working class. Numerous middle-class environmental organizations—NIMBY movements—have formed to oppose the siting of hazardous facilities or other perceived threats to health or property posed by environmental degradation. Although I am unaware of any thorough account of organizations formed to oppose immediate environmental threats, it is likely that there are more such organizations composed of middle-class constituents than those with working-class members. Indeed the significant amount of scholarly attention received by the environmental-justice movement is more a reflection of the relative rarity of working-class mobilization around such issues compared to that of middle-class communities where more abundant financial and other resources allow for a generally higher level of political action.

But even if it should turn out that interest-oriented environmental mobilization is more common among the working class, this can more readily be understood based on the fact that working-class residents face far more immediate environmental threats than do those in the middle class. Numerous studies have demonstrated that the working class endures a disproportionate share of the cost in terms of health and safety that results from environmental hazards (Bryant 1995; Bullard 1993; Lester, Allen, and Hill 2001; Westra and Wenz 1995). The siting of hazardous facilities systematically corresponds with community income levels. The point here is that interest-based mobilization around environmental issues is not unique to the working class. The working class may act more often in regard to immediate environmental issues simply because its interests are more often threatened by environmental hazards. Relatively privileged middle-class citizens find themselves called upon to mobilize less often against threats to their interests, because they have the political power to deter such threats. Yet when a threat does materialize, they are no less inclined, and possibly even more inclined, to mobilize in defense of their interests than are working-class people.

This still leaves the question of whether or not the organizational structure and practices of working-class political movements outside of organized labor differ from their middle-class counterparts. In terms of their organizational form, there is little evidence to suggest that voluntary working-class SMOs differ significantly from voluntary middle-class SMOs. When we examine nonprofit advocacy organizations, several of the characteristics attributed to class culture can be more easily understood by looking at the structural constraints that shape them.

Rose (2000) offers the following class-cultural explanation of middle-class political form:

Middle-class organizations are designed to accommodate individualism and even encourage it as a fundamental value. They tend to emphasize egalitarian decision making and allow people to express their own ideas and evaluate arguments for themselves. Attention to group process is important for allowing individuals time for self expression. . . . Membership implies no particular responsibilities or duties, and individuals are primarily beholden to their own conscience or sense of responsibility. . . . Middle class politics is therefore an extension of personal development. (67)

Yet the characteristics of middle-class organizations that Rose describes can be expected in any organization that relies upon voluntary support and contributions. Members of such organizations are "beholden to their own conscience or sense of responsibility" because, by the very nature of voluntary organizations, there is no other force compelling them to participate. Regardless of members' class backgrounds, this is likely to yield more flexibility in terms of what is expected of them. An authoritarian hierarchy and rigid rules of participation will deter voluntary supporters, regardless of their economic class. If some concrete return is expected from membership, as in the case of workplace representation, and if there is a monopoly on representation, as there is with unions, members may be more tolerant of hierarchy and rigidity in terms of organizational procedures. When voluntarism is instead the basis of participation in the organization, the amount of procedural rigidity and hierarchy in the organization will decline, regardless of the economic class to which its members belong.

But even given the tendency to encounter more flexibility and egalitarianism in voluntary organizations, the claims of class culture and New Social Movement theorists regarding the "internal modes of action" of middle-class SMOs are overstated. According to Offe (1987), new social movements "have at best rudimentary membership roles, programs, platforms, representatives, officials, staffs, and membership dues" (70). This may be true of some organizations on the fringe, such as Earth First!, but it hardly characterizes the professional environmental SMOs that represent hundreds of thousands of environmental supporters in the United States (Gottlieb 1993). These organizations have clearly established platforms and programs, a hierarchy of elected representatives and officials, professional staffs, and systematically monitored dues payment requirements for members.

The formality or informality of SMOs has more to do with their stage of development than to any class-cultural predisposition of the membership. Suzanne Staggenborg (1988) identified a process of professionalization and formalization that occurs over the life of movement organizations. Beyond the legal and structural mechanisms that tend to generate greater formality in the labor movement, labor unions are also much older than most of the new social-movement organizations. But even the older and more established environmental groups already have a more formal organizational structure, despite their largely middle-class membership.

What's more, some research suggests that participants in these professional movement organizations are drawn to them specifically *because* of their formality and the political power that they believe comes with it. Some report joining so that they can receive information that tells them what they should do to support the cause (Obach 1994). Far from being expressive, participatory outlets, most professional SMOs are structured in such a way that a few central leaders conduct analysis and make decisions and then inform and instruct an otherwise inactive membership. This contradicts claims about the desire of middle-class activists for participatory involvement and political self-expression. Although they are perhaps not as rigid in their organizational form and procedures as labor unions, it is clear by looking at these professional middle-class SMOs that distinctions between middle-class and working-class organizations based on form and structure have been drawn too sharply.

This critique applies as well to Offe's claims regarding "external modes of action," the way NSMs "confront the outside world and their political opponents" (Offe 1987, 70). Offe emphasizes their reliance on demonstrations and other unconventional tactics. New social movement organizations may more often engage in protest politics than labor unions. As demonstrated by the Earth First! example offered at the beginning of the chapter, the antics of some new social movement activists may clash with the norms of some working-class unionists. Professional lobbying, however, makes up a good portion of environmental-movement activity. In addition, as the events of 1999 in Seattle demonstrate, labor unions make use of protest politics as well. Picketing, marches, and sit-ins were invented by the labor movement and have existed within the political repertoire of unions for over one hundred years, long before environmental SMOs adopted such tactics. Again, the distinctions asserted to exist between middle- and working-class organizations by class culture and New Social Movement theorists are better attributed to specific features of organizations and their positions within the political and social structure.

There is also the question of the shifting priorities identified by class culture and New Social Movement theorists as characteristic of middle-class organizations. According to the class-cultural explanation, this propensity to shift priorities readily can be attributed to the fact that these organizations select issues not on the basis of the relatively stable *interests* of members, but rather, on the basis of what participants find interes*ting*. A class-cultural analysis holds that political activity not only provides a sense of fulfillment, but also an opportunity for exploration and growth. Middle-class activists are thought to align voluntarily with one cause and then to withdraw from participation and find their way into another organization or to redirect the priorities of the one they are in, according to their sense of whether these needs for fulfillment and exploration are being met. Whereas working-class activists remain tied to the same workplace or community-based interest organizations, middle-class activists drift from one issue to the next as their values or interests shift.

Again, this explanation is consistent with some of the evidence found in the cases examined in this study. As indicated above, several of the labor leaders interviewed noted the fluctuating concerns of environmentalists and expressed frustration in their attempts to develop a common long-term program. As noted in chapter 5, however, labor organizations have a stable funding base and a monopoly on representation of their members. Environmental and other voluntary movement organizations, on the other hand, have to constantly seek ways of attracting and retaining interest, members, and funds. Since they do not have any kind of legal monopoly on representing a given cause, they also find themselves in competition with numerous others doing similar work. The competitive quest for recognition that results leads environmental organizations to shift priorities more often than labor unions do. According to Rick Isaac of the New Jersey Sierra Club, "it's a hustle to get the public's attention, to find the issue that plays best" (personal communication, May 24, 1999). This places environmental organizations in a situation in which they may be unable to pursue a particular issue in an ongoing way but instead must seek issues that have public appeal at a particular time.[1] The tendency to shift priorities is further encouraged by rules that limit the size of the asset base that nonprofit movement organizations can accumulate. In their analysis of the channeling mechanisms that shape the organizational behavior of nonprofit movement organizations, McCarthy, Britt, and Wolfson (1991) report that "[r]etaining assets in order to guarantee long term organizational stability . . . is not acceptable. Adhering to this rule makes SMOs dependent on short term markets for their social change products" (67).

In addition to the monetary support they receive from individual members, many environmental organizations also rely on funding from foundations. In many cases the funding that foundations provide is tied to specific uses and cannot be used to aid the organization in its routine functioning or to fund ongoing efforts. Instead foundation funds are commonly dedicated for specific finite projects at the end of which organizational staff members must have another project proposal ready to go. The competitive aspect of this work requires that these organizations constantly develop appeals not only for new members, but also for foundation support. According to Barnett of the Healthy Schools Network, this also contributes to an ongoing quest for attractive new issues: "There is a fair degree of competition among environmental organizations in New York state and elsewhere, because they are all competing for foundation funding—Who's got the best issue? Who's got the best definition? Who's got the best quotes in the media?" (personal communication, May 12, 1999). The comparatively stable programs adopted by unions are a product not only of their legally defined interest-based mission, but also of their unique structural position, which provides them with a secure membership base and resource flow. The shifting priorities of middle-class environmental organizations can likewise be understood in terms of their constant need to attract members and funds, rather than being seen as reflecting the fluctuating concerns of activists seeking means of self-expression and personal development.

An additional structural limitation faced by most environmental organizations is the legal prohibition on their participation in partisan politics. As nonprofit organizations falling under Title 26, Section 501(c) (3) of the Internal Revenue Code, they are not allowed to endorse candidates, make campaign contributions, or otherwise engage in electoral activity (Bucholtz 1998; McCarthy, Britt, and Wolfson 1991). This prohibition eliminates a vast area of potential cooperative activity between unions and environmental organizations and heightens perceived differences in their general approach to social change. According to Evelyn DeFrees of the Natural Resources Council of Maine,

The labor community is totally wrapped up in political process. . . . I do think that there is traditionally a view . . . that environmentalists, along with the [perception of a] class-based elitism, also keep themselves above the fray. They don't dirty their hands with the races, who wins, and that sort of stuff; that they are pure issue-oriented people. Whereas, labor folks go out and scrap away and make the thing happen. There is more of a rough and ready aspect to it, as opposed to this almost academic, sit in your office and think of what is the right policy option. . . . Environmentalists are very slow to get involved in politics, and environmentalists are, in general, nonprofit organizations. (personal communication, May 20, 1999)

Jeff Jones of Environmental Advocates in New York identifies the way in which the legally mandated nonpartisan nature of environmental organizations yields a different approach to politics than that of labor unions:

The big difference between us and [the unions] is that we don't support candidates and we don't make contributions to candidates. Our power is entirely the power of public opinion. . . . Since unions endorse candidates, the unions around here tend to endorse incumbents. The rate of incumbency in New York state is over 98 percent. You can almost never be defeated if you are a member of the Senate or Assembly. The only way you ever get in is if someone runs for higher office or dies. Institutions like labor have decided long ago that they don't get very much out of backing political challengers. So you go to an endorsement meeting, like I have, at the AFL-CIO, and they are down the line just endorsing the incumbents, one after the other. So it's a very different approach to politics than our approach. (personal communication, May 7, 1999)

Again, we see a very different type of politics taking place among unions and environmental organizations; the difference, however, is best understood not in terms of class culture, but in terms of the structural limitations on the organizations involved. Partisan politics involves some amount of deal making and trade-offs and an overall different strategic assessment than the politics of nonpartisan organizations that are required to remain "above the fray," taking principled positions exclusively on issue campaigns.

Of course, some environmental groups forgo their nonprofit status and do engage in electoral politics. When they do, their candidate endorsements tend to parallel those of organized labor. In some cases, this serves as a basis for cooperation between unions and environmentalists, again demonstrating that when structural barriers are removed and interests align, cultural differences do little to inhibit collaborative efforts.

The evidence presented here suggests that class culture is an inadequate explanation for the differences between labor and environmental organizations reported by coalition participants. The legal and structural constraints on the operation of these organizations better explain these differences. To summarize, the interest-based orientation of labor unions is found primarily at the local level, where unions are legally required to represent the workplace interests of the employees who make up their membership. As we move to higher levels of organization within the labor movement, where legal mandates do not restrict organizational goals, we find more in the way of values-based political activity. Legal requirements also facilitate the imposition of a hierarchical structure and majority rule decision making within labor organizations. Voluntary organizations face no such requirements, and given that they have no means to coerce support, greater organizational flexibility and greater accommodation of their volunteer members is the logical result. Whereas labor unions with their legal mandates and stable funding base more often focus on member interests, voluntary organizations must, in an ongoing way, strive to identify new issues to attract members and foundation funding. The legal prohibition on their participation in electoral politics also leads nonprofit organizations to focus purely on issues, whereas labor unions, free to engage in partisan politics, enter the realm of deal making and trade-offs associated with the electoral arena. In each of these two types of organization, structural variables can be seen as defining their form and practices. Despite the overriding effect of these structural variables, many cases demonstrate contradictory tendencies, with, for example, professional environmental organizations taking on hierarchical forms or labor unions assuming values-based positions. Each of the behaviors and organizational forms attributed to class culture and the distinct production practices of middle- and working-class actors can be understood more simply in terms of the structural limitations placed on these organizations.

This is not to suggest that class culture plays no role in explaining these difference. In terms of understanding intermovement relations, however, these different explanations offer very contrasting predictions about when and under what circumstances cooperation or conflict will emerge between working- and middle-class organizations. Differences embedded in class culture suggest that a significant barrier between working- and middle-class SMOs will be found in any intermovement relationship. Differences rooted in the structural limitations placed on organizations, however, suggest that those in similar structural positions can engage in cooperative relations with little difficulty, regardless of their class makeup. And some organizations, while differing in class makeup, occupy similar structural locations. According to such a structural analysis, cultural differences will pose few problems for these organizations. This point will be addressed in greater detail in the next section.

One issue that has yet to be addressed is the disparity between working- and middle-class participation in voluntary social-movement organizations. The structural analysis offered above may explain why some differences may be found in the form and functioning of unions and voluntary movement organizations; it does not, however, explain why members of the middle class tend to join voluntary organizations with much greater frequency than their working-class counterparts. Certain elements of the cultural explanations offered by Rose and others may have validity here. Middle-class professionals may seek expressive forms of political involvement in a way that members of the working class do not. However, equally plausible alternatives should also be given consideration.

There may be class differences in terms of how social-change activity is engaged, unrelated to the theorized interest foundation of working-class activism. For example, it is possible that working-class volunteers, acting on their values as opposed to their private interests, may more often engage in charitable work as opposed to political action. It could also be the case that members of the working class more often undertake political activity through religious organizations and thus do not register in most analyses of SMO participation. Or it may simply be that voluntary organizations that rely on financial contributions more often target middle- or upper-middle-class communities in their membership solicitation efforts, based on the idea that people in those communities have more disposable income or time to donate than those in working-class communities, a possibility that in itself may explain differential participation. SMOs that rely upon canvassing or phone solicitations as their primary means of recruiting supporters target "high probability contributors" (Oliver and Marwell 1992). Their lists of potential contributors are likely to include a class- and social-network dimension that would preclude widespread recruitment of working-class constituents into mainstream environmental organizations. Research on networks among movement organizations has found few ties between middle-class environmental SMOs and unions or other working-class movement organizations (Carroll and Ratner 1996; Salisbury et al. 1987). Members of the working class may be underrepresented in environmental SMOs for reasons as simple as never having been asked to join or, if they have been asked, simply lacking the disposable resources to pay the required dues. Basic organizational practices and social-network issues such as these may explain why the middle class is disproportionately represented in environmental-movement organizations and other voluntary advocacy groups, and each of these potential explanations needs further exploration before claims of broad cultural differences are accepted as the reason for differential patterns of participation in environmental organizations.

Do Differences Represent Barriers?

This discussion has assumed throughout that the differences identified between labor and environmental organizations in organizational functioning do in fact serve as barriers to intermovement cooperation, whether the differences are rooted in class-culture or in organizational or structural forces. Indeed, several labor and environmental leaders interviewed in this study did identify difficulties related to some of the issues raised by class culture theorists, and some of these differences are best considered to be cultural in nature. On the whole, however, these difficulties were not considered by many of the subjects interviewed to be significant problems standing in the way of alliance formation. On the contrary, although most of the leaders interviewed were able to identify some differences in the practices of one another's organizations (and a significant minority were not able to identify any real differences at all), none suggested that these differences lay at the root of their inability to work cooperatively with organizations from the other movement. Other than a few minor missteps, subjects reported that they readily picked up on the different organizational styles of alliance partners from other sectors and were able to adjust their own procedures to bring them into alignment with those of the organizations with whom they were cooperating. Environmentalists discovered that they had to call the leader of a union to set up a meeting rather than a rank-and-file activist. Labor leaders learned to tolerate or redirect discussions when an environmental activist began theorizing about the larger purpose of the movement. Some environmentalists even reported picking up the "brother" and "sister" nomenclature commonly used to address fellow participants in labor meetings. In short, whatever cultural differences may exist between these two types of organizations, participants in coalition activity did not see them as significant. The fact that even significant cultural differences can be overcome in the context of political struggle has been found in other cases as well (Bevacqua 2001; Grossman 2001; Lyons 2001).

One possible explanation for the fact that existing cultural differences thought to be significant by other researchers were not found to be important in this study has to do with the organizational level at which this research project was conducted. Many of Fred Rose's conclusions are based on surveys, interviews, and observation of rank-and-file workers and grassroots peace and environmental activists. For this project, I conducted interviews primarily with state and national labor and environmental leaders. Thus, the level of focus may explain the lack of cultural barriers found in this study. Although I gathered no systematic data on the class backgrounds of the organizational leaders who were my subjects, anecdotal evidence suggests that at least in their present circumstances, there are few cultural differences between them. Except for a few volunteers, most make their living working for medium- to large-sized membership-based organizations. All are engaged in political advocacy and have at least some experience working within the halls of Congress or the state legislature. Their incomes may vary somewhat, but all probably fall within the middle income brackets. All of the subjects appeared to be white except for one lobbyist known to be Latina.

Based on their language and writing skills, most appeared to be well educated. Socially, they travel in similar circles. One, a political director of the New York State AFL-CIO, is married to an environmental lobbyist. Another, the secretary treasurer of the Maine Federation of Labor goes canoeing with a staff member of the Natural Resources Council of Maine. The former executive director of a large national environmental organization is married to the president of an international union.

These social ties and occupational similarities among the interview subjects suggest that cultural differences are not significant at the level of organizations that I studied. More significant cultural differences might be found between a grassroots member of the Audubon Society and a rank-and-file member of the carpenters union. This is not, however, the level at which intermovement coalitions are formed. The class-cultural distinctions identified between the working and middle classes may be valid but misspecified as a basis for attempting to understand relations between the *leaders* of labor and environmental organizations. If these leaders and activists are at all different from rank-and-file members in terms of their political understandings (and their involvement as leaders or activists alone suggests that they are), then the general cultural framework within which they operate may not be that important in explaining the dynamics of coalition participation. They have political sensibilities that already distinguish them from the nonactive majority, and thus it is their attributes, not those of the membership or their class as a whole, that are of crucial importance in understanding coalition formation.

Rose talks about movement organizations as "schools of democracy" in which ordinary citizens achieve empowerment and attain a political education. Movement coalitions, especially interclass movement coalitions, he argues, provide citizens with a greater sense of understanding of and a shared sense of purpose with people of different backgrounds. As discussed in chapter 6, I also found the educative function of coalitions to be important. Yet although mass grassroots participation in coalition activity may be the ideal, the reality is that coalition work is typically carried out by a handful of organizational leaders. Rose (2000) himself acknowledges that the successful intermovement coalitions he examined "were the work of a very small number of activists" (184).

For the most part, the same is true of the coalitions found among the movements in this study. In their ideal form, movements may be grassroots based, democratic, and participatory. On occasion such movements do arise. For better or worse, however, other than these periodic "waves of protest," movements are carried on by a small core of activists and professional movement organization leaders.

The success of a coalition campaign does depend in part on the extent to which the grassroots base of the associated organizations can be mobilized (Nissen 1995). But as described in chapter 6, the education of the grassroots membership is often carried out by the leaders of the organizations to which they belong. Rarely does this education take place through a process of widespread direct interaction among diverse grassroots members. Thousands of "Teamsters and Turtles" marched together through the streets of Seattle in 1999, but for the vast majority, those few hours represented the full extent of their direct contact with one another. A handful of leaders and activists built the coalition and successfully mobilized their respective bases to great effect. Thus significant cultural differences that may or may not exist between the grassroots members of different organizations will seldom come into play in determining whether and how coalition participation occurs. It is the leadership of the organizations participating in a coalition that is of most importance in understanding coalition processes. Based on the data gathered in this study, to the extent that cultural differences exist at all between leaders in different social movements, they present only nominal barriers to intermovement cooperation.

Other Cultural Barriers?

Thus far this chapter has focused on the question of class culture and whether or not theorized class-cultural differences serve as a barrier to intermovement cooperation. One could raise similar questions regarding cultural differences in general (Bystydzienski and Schacht 2001). The growing ethnic diversity in the United States is, to a degree, reflected in the social-movement organizations that populate the political landscape. One might argue that the cultural differences that arise from this growing diversity could contribute, or already have contributed, to tensions between social movement organizations. A number of inter-movement conflicts could be interpreted in this light. For example, some environmental-justice organizations composed primarily of people of color have been very critical of the primarily white mainstream environmental movement for their failure to address health hazards; in the 1970s there was open conflict between the primarily Latino United Farm Workers Union and the predominantly white Teamsters union; some African American civil rights organizations have also been critical of white environmental organizations for advocating policies that they believe threaten economic opportunity in minority communities; in the late 1990s in Los Angeles, Latino landscapers came into conflict with white homeowners over the use of loud, polluting leaf-blowing equipment. Each of these intermovement conflicts has a clear cultural dimension. Although cultural differences are not necessarily at the center of these conflicts, those differences may prevent common understandings and the resolution of problems.

The growth of the environmental-justice movement and the increasingly important role that an expanding Latino population is playing within the labor movement indicate that ethnic differences in social-movement politics must be given further consideration. The styles and cultural referents used by activists and leaders of different ethnic backgrounds may prove to be important in understanding how they approach political issues and how the SMOs they head relate to one another. There is evidence to suggest that environmental issues in particular are viewed through distinct ethnic lenses. In his study of environmental action among different ethnic groups in Los Angeles, Robert Gottlieb (2001) notes that important cultural differences among these groups must be taken into consideration regarding the way they approach environmental issues.

But it is as yet unclear whether or to what extent such differences serve as barriers to intermovement cooperation. In the instances cited above regarding tensions between SMOs of different ethnic makeups, it is not clear whether cultural differences should be seen as fundamental to the problem. Clearly different groups have different needs and interests, given their structural location in the society. For example, new immigrants or members of an ethnic group that has experienced discrimination resulting in relatively high poverty rates may prioritize economic concerns over environmental ones. This prioritization may result in conflict with an environmental group that opposes a potentially job-generating, but environmentally hazardous, project. Such a basic clash of interests is likely to lead to conflict, but it would probably do so regardless of the cultural differences between group members. An impoverished white community group would likely clash with environmentalists on such an issue as well. One must be careful to distinguish between conflicts that are rooted in cultural difference and conflicts between groups of different cultural backgrounds based on issues that are not necessarily tied to culture. Making such distinctions is challenging, since the history of racism in the United States has created populations that are both culturally distinct and disproportionately relegated to certain disadvantaged structural positions in U.S. society. Assessing whether cultural differences or basic interests lie at the root of a particular intermovement conflict can therefore be quite difficult.

Yet there is some evidence to suggest that even very deep cultural rifts between groups and between individuals can be overcome in the course of political struggle. Zoltan Grossman (2001) examined the emergence of an interethnic coalition among Native Americans and rural white people in northern Wisconsin. Deep divisions between Native Americans and white sport fishers arose during the late 1980s and early 1990s over the issue of Native fishing rights. The conflict between them reached the point of violent confrontation with explicit racial overtones. Yet these two communities were able to overcome differences and work together when fishing waters were threatened by a proposed mining operation. These two groups formed a successful environmental alliance to defeat the mines. A similar pattern can be found in several other instances in which Native Americans and rural white communities overcame differences to form an effective coalition. According to Grossman, several elements make up the foundation of such interethnic alliances:

First, alliances have emerged from a "sense of common place," reflecting the participants cultural ties to the local geographic landscape and a cultural bond to the environment. Second, the alliances have been formed by a "sense of common purpose" or the idea that Native Americans and rural whites are "in it together" in legal, political and economic terms. Third, the alliances are sometimes born out of a "sense of common understanding" or a conscious determination to over-come the ethnic groups' conflict by building mutual goals. (154)

Geographic circumstances played an important role in the interethnic alliances cited by Grossman, allowing for a sense of common place and purpose to exist. But it is likely that any threat to common interests can help distinct ethnic groups to overcome cultural differences, in a way similar to that in which common interest in environmental protection has made it possible for middle-class environmentalists to join with blue-collar workers.

It is impossible to offer a firm prediction regarding the issues involved in the formation of interethnic alliances based on the data gathered in this study. As noted earlier, the leaders and activists of environmental and labor organizations studied here are a relatively homogenous group ethnically. But based upon the findings in this study regarding class-cultural differences and those of researchers examining the ethnic-culture question, one would predict that cultural differences based on ethnicity should not serve as a serious impediment to intermovement cooperation, provided the interests of the groups involved are not directly opposed and coalition brokers are present to aid in the development of mutual understandings. Those who come to hold leadership positions in politi-cal organizations commonly develop sensibilities that enable them to interact with people of different backgrounds. Although they may lead an organization with unique goals and a distinct membership, their role does not differ significantly from the leaders of other political organiza-tions. Working with other movement organizations may require some sensitivity and effort to adjust behavior according to differences in style, ideology, and political approach, but if common goals can be identified, it is unlikely that cultural differences will present a fundamental barrier to intermovement cooperation.

Conclusion

Differences in culture offer a tempting explanation for variations in the behavior of groups or individuals that are otherwise not readily under-standable. In some research in this area, theorists treat culture as a free-standing independent variable, whereas in more sophisticated treatments, theorists examine the historical or structural forces that shape culture. In the research reviewed here, some scholars have insightfully grounded cultural distinctions regarding the motivations and practices of different movement organizations in class and the production process. These scholars believe the interest-driven basis of working-class politics and the corresponding hierarchy and centralization found within their movement organizations reflect manual-labor practices and the daily conflicts with supervisory authorities that characterize blue-collar work life. In contrast, the values-driven, participatory form of middle-class political involvement is thought to mirror the self-motivated, expressive work life of the professional. Although the cultural theory on which these explanations are based is well grounded in the structure of work life, the explanations may not be as compelling as those advanced by a less abstract structural account as to why we see practical differences between working- and middle-class movement organizations. One need not examine work structures to identify class-cultural differences that then find their expression in organizational functioning if there are readily apparent structural factors that directly influence organizational functioning.

Although the theoretical basis for the arguments put forth by cultural theorists is intriguing, there are several issues that call its validity into question. The cases selected for examination by some cultural theorists interested in intermovement conflict present the first problem. The dramatic conflicts surrounding forest issues may attract the most attention of any of the struggles between labor and environmental forces, but in many ways the contrast between timber industry workers and environmentalists is a poor representation of interclass differences. Timber industry workers are unique in their relationship to nature, the contingent structure of their work life, and their geographic isolation. In many ways they represent the most extreme contrast between workers and middle-class environmentalists and in no way serve as a good representative case for a general analysis of class-cultural distinctions. The use of unions in general as a representative basis for analyzing working-class political form is in itself problematic. Legal mandate determines the interest focus of collective bargaining and grievance procedures as well as their hierarchical organizational form. Thus these characteristics of unions are better attributed to law than to working-class culture. This becomes more apparent when we consider tendencies that contradict these general union characteristics, such as the values-driven goals of union federations free from the constraints of day-to-day workplace struggles or the public-goods focus of working-class unions operating under different structural constraints in other nations.

The attributes of middle-class political form and practice are also more easily explained by looking at the structural conditions under which they express themselves. Lacking a secure funding base, voluntary SMOs are forced continually to seek attention-grabbing issues in order to attract members and foundation grant money. This more likely reflects financial necessity than an inherently fickle, expressive nature of middle-class professional culture. In addition, the greater flexibility and informality associated with environmental SMOs is better considered a product of their voluntary nature rather than a cultural attribute. No laws specify decision-making procedures or rules of representation for environmental groups. And unlike unionists, participants in these organizations are not compelled to join, nor can they expect much in the way of material return for their contribution. In light of this, participants are less likely to tolerate rigid hierarchy or stifling formality, a predisposition that could probably be found among volunteers of any class. But even given these legal and structural factors, the stark cultural contrasts asserted by cultural theorists to exist between the two types of organizations are not always evident. Many professional environmental SMOs, for example, do have a considerable degree of formality and hierarchy despite their middle-class membership base.

At some point behavior dictated by legal or structural mechanisms may become so ingrained in an organization's practices and the belief system of its members that it is most appropriately conceived of in terms of culture. It has then become a characteristic of the organization that will continue indefinitely even if the original constraint that created the characteristic is removed. As long as the constraint that requires certain organizational practices is still in place, however, it is inappropriate to refer to culture as the causal mechanism for those practices. In this chapter I have attempted to demonstrate that many of the characteristics of unions and environmental SMOs attributed by some theorists to class culture are more properly considered to be reflective of legal and structural constraints on those organizations. At some organizational levels these constraints facilitate practices and orientations that serve as a barrier to intermovement cooperation. Yet at other levels, the structural location of unions and environmental organizations is not that far apart. If we consider union federations, as opposed to locals or particular internationals, we see organizations that are more similarly situated to professional environmental SMOs. At this organizational level, we see relatively unproblematic cooperation between unions and environmental organizations when interests align and the necessary personnel are present to bring the organizations together. But even to the extent that divisive class cultures exist, they were not found to be significant among the individuals who lead these organizations. It is they who must forge the interorganizational bonds that allow for political coalition activity, and there is little evidence of a significant cultural divide between them.

9

"We Can Work Together": Uniting the Pieces

Whereas most movement scholars have chosen to focus on either microlevel issues concerning how individuals are drawn into movement participation or on macrolevel questions regarding when mass mobilization of movement participants will occur, relatively little attention has been paid to movement politics at the organizational level (Zald and McCarthy 1980). This is surprising given that organizations have long been recognized as central to social movement activity. Not only do they mobilize citizens, but they also define goals and the strategies necessary to achieve them. SMOs play a key role in articulating and advancing the policies that ultimately represent the societal changes that movements exist to achieve. Yet given the fragmentation that characterizes social movement politics in the United States, it is coalitions, and not individual movement organizations, that are most capable of affecting public policy and the general direction of social change. That being the case, it is essential that more research be directed toward the study of coalitions among social movement organizations. This book explores this crucial strategic issue. Using two of the most important movement sectors in America, I examine how relations between them determine whether they will form coalitions or whether they will fail to do so and instead align with others in opposing camps. I began the book with the observation that unions and environmental SMOs have both cooperative and conflictual relations. This variation in their relations provides the ideal context for elucidating the factors that determine whether or not they will enter into political alliance. In each of the subsequent chapters I explored the structures and processes that affect these relations. In this final chapter I will review and synthesize the work presented in previous chapters and draw out the larger theoretical implications of the patterns I have identified.

Most coalition theorists begin with basic assumptions about the goals that organizations are striving to achieve and the strategic calculations that go into decisions regarding coalition activity. Their basic premise is that organizations will agree to share resources and coordinate action when doing so will better enable them to achieve their goals and will refrain from coalition activity when it is not necessary to achieve their ends or if their potential partners are pursuing goals counter to their interests. In examining labor-environmental coalitions, the key question from this perspective is whether environmental goals will have a positive or negative impact on the economic and workplace interests of workers.

Chapter 2 addressed the impact of environmental measures on the core interests of workers. The evidence on this issue suggests that environmental measures on the whole have a positive economic impact and can create jobs in many instances. Although the broad economic gains associated with environmental protection have not led to unconditional union support for environmental measures, when particular job gains are apparent, unions have supported environmental measures. But generally, many in the labor movement remain skeptical about any positive economic impact of environmental protection. This skepticism has been fueled by high-profile cases in which environmental measures have had negative job consequences for certain workers. Such job losses are concentrated in the extractive industries, in particular, the coal and timber sectors. Conflict between environmentalists and workers in these sectors has received a great deal of media attention despite the rarity of this kind of conflict and the relatively small number of jobs that have actually been lost.

But despite the attention that these cases have received and the skepticism it has engendered on the part of unions in regard to positive economic impacts of environmentalism, the kind of hostile relations evident in these cases are in no way characteristic of labor-environmental relations in general. On the contrary, unions and environmentalists have found themselves on the same side of many issues, including those related to workplace health and safety, right-to-know laws, and international trade, in addition to their close alignment in electoral politics. Although conflicts between the two groups do occasionally crop up, on the whole relations between unions and environmentalists are far better than indicated by popular media accounts or scholarly research. Although it is still best to conceive of unions as being situated between the primary rivals in environmental conflict, environmentalists and industry, in the political sphere it is more common for unions to cooperate with environmentalists than to be in conflict with them.

Economic analyses provide some insight into the nature of the relationship between unions and environmental-movement organizations, but an examination of interest convergence or lack thereof reveals only part of the story. The second half of this research project used case studies to pick up where the core-interest model developed in the first part of the book left off. Rather than engaging in a simple interest analysis based on an assessment of the static goals of movement organizations, it is necessary to examine the structural conditions that shape the goals sought by different groups and the dynamic nature of interorganizational relations. Chapter 3 provided a historical overview of labor-environmental relations in the United States. It is clear from this historical overview that economic forces and other factors that influence the core interests of the relevant actors have had some effect on the way in which these two movement sectors have interacted. But it is equally clear that many other variables also come into play and that the quality of labor-environmental relations cannot be determined based solely on the way environmental policies influence jobs or any other presumed interest. A closer look at intermovement processes in several states served as the basis for understanding each element in the development of this relationship. After an overview of each state examined was provided, these elements were analyzed in detail in chapters 5 to 8. Although it would be impossible to incorporate every contingency into a single model of coalition formation, here I will attempt to assemble the various pieces explored in this part of the study into a more complete picture of when and how coalitions form. This provides us with a general framework for examining the relevant determinants of coalition formation, from the broadest structural forces shaping the behavior of social groups to the personal interactions of individual coalition brokers.

A Synthesis of Coalition Tendencies

In chapter 5 I began by analyzing the basic structural conditions that give rise to different movement organizations and how the position of organizations within that structure shapes their ability to join in coalition with others. The structural foundation of coalition patterns in a particular society can be found in the political and economic institutions of that society. Although many theorists suggest that patterns of coalition formation can be identified among any groups or individuals, political structures determine the behavior of SMOs regarding coalition formation. The most important characteristic of the U.S. political system in this regard is the nonunitary system of government, which facilitates the development of multiple competing movement organizations (see figure 9.1). Labor unions and environmental SMOs are among the many types of organizations to emerge out of this political structure.

The tendency in the political climate created by the U.S. political system toward issue and organizational fragmentation can clearly be seen in the case of the environmental movement. Lacking any official recognition or mechanism for obtaining it, numerous groups have formed around a myriad of particular environmental issues. The labor movement is unique among the popular-interest sectors generated by the U.S. political structure because it has been given a level of formal government recognition. Whereas organizations from other movement sectors are all governed by a minimum set of guidelines, mostly associated with tax status, labor unions are more highly regulated and operate under specific laws governing their form and behavior. Although still fragmented in many ways, the labor movement is characterized by a more centralized configuration relative to that of other movement sectors, demonstrating the significance of political structures in shaping SMO form and behavior.

Most, if not all, political conflicts in a society are also significantly shaped by its economic institutions. Many of the movement organizations that populate the political landscape are formed on the basis of economic interests or at least engage in disputes over the distribution or control of public or private resources. American welfare capitalism, along with its regulatory and redistributive apparatuses, forms the economic context in which conflict or coalition building among movement organizations will occur. Labor and environmental-movement organizations can be located within this institutional structure. Unions are organized primarily around redistributive issues, as exhibited both in workplace bargaining and in the goals they strive to achieve through political action. Union efforts have a regulatory dimension as well, however, in terms of workplace safety and health and other employment practices.

Environmentalists, on the other hand, are focused primarily on the regulatory dimension of economic policy. They oftentimes aim to restrict, for the sake of environmental protection, the use of private property and to expand the base of publicly controlled and protected resources. Their redistributive and regulatory goals within the U.S. economic structure creates the context for labor-environmental relations. Within that context, contradictory tendencies emerge. The potential for alignment between the labor and environmental sectors is apparent in the support of both for the regulation of private industry, and in that sense they share a common adversary. In some cases the range adopted by organizations within each sector overlaps, as when organizations within the two sectors seek to replace production practices that pose threats to both worker health and the environment. But in other cases their support for government regulation diverges, as when workers sense that environmental policies will threaten industry profitability to the extent that their own livelihoods may be imperiled. Thus the political and economic institutional arrangements in the United States clearly create the context in which movement organizations will form, which in turn influences the range of issues that these organizations undertake to address and how they will relate to one another within this context.

The basic institutional context within which movement organizations in the United States operate was established at the founding of the nation; it was not until certain legal and social changes occurred, however, that we see the engagement of labor and environmental actors that forms the basis of this study. Although the nation's founders allegedly designed the government structure to disperse power and prevent influential factions from exercising undue control, for much of American history, policy was shaped largely by a small number of elite groups. Popular-interest groups

**Nonunitary
System of
Government**

**Welfare
Capitalism**

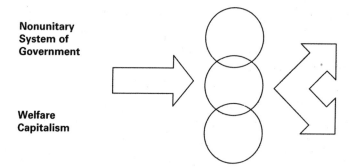

(a)
The nonunitary system of government generates multiple competing SMOs, many, if not all, of which are greatly shaped by the economic structures that determine the distribution of and control over resources. The U.S. system of welfare capitalism, characterized by market functioning along with redistributive and regulatory interventions, creates the framework within which SMOs form and compete politically.

(b)
When their numbers were fewer, organizations could maintain a broad agenda with some inter-organizational overlap, while still remaining distinct. However, beginning in the 1960s, government expansion, increasing resource pools, and other factors contributed to a proliferation of movement organizations. The circles represent distinct organizations; the range of issues those organizations address is symbolized by the diameter of the circles. The solid line surrounding the circles denotes their distinctive identity.

Figure 9.1

have existed since the founding of the nation; until the second half of the last century, however, they were relatively few in number and limited in their influence. Occasional waves of protest and various social crises enabled these groups to have a significant impact on government policy at certain times; routine policy formation during "normal" times, however, was largely the province of economic elites closely tied to political leaders.

An explosion of social movement activity and a proliferation of social movement organizations during the 1960s, disrupted, to some extent at least, the insulated control of policy by elites. In a synergistic spiral, government programs expanded, and new populations served by those

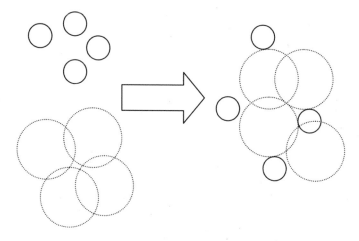

(c)
The "coalition contradiction": Given a more crowded field, organizations are compelled either to narrow their range to clearly differentiate themselves from competitors or to maintain a range broad enough to engage in coalition work to increase efficacy. Although a broader range may better enable an organization to work with others to achieve its goals, it also makes it less distinguishable in its efforts to win supporters.

(d)
Some organizations adopt different strategies, allowing them to engage in coalition work, whereas others maintain a more narrow range, allowing for relatively less coalition participation. Note that an overlap in two organizations' range does not mean that a coalition will necessarily form, only that they have a greater potential for coalition formation. Other conditions and organizational characteristics must also be present for a coalition to form even when an overlap in range exists. It is also possible for an organization with a narrow range to engage in coalition activity, although this activity is more commonly of a purely instrumental type.

Figure 9.1
(continued)

programs organized to foster that expansion. Increasing resource pools in the prosperous postwar economy, higher levels of educational attainment among the populace, government reform, and other changes all contributed to the political landscape's becoming more densely populated by interest groups, including popular social-movement organizations capable of posing a greater challenge to elite policy insiders.

Established government structures in the United States have always tended toward the fragmentation of interests, and the expanding welfare state further enhanced that tendency. Separate agencies and legislative committees created narrow policy realms around which multiple movement organizations would form. The more crowded political landscape generated by the explosion in movement activity magnified this issue fragmentation. Given their relatively small numbers, early popular movements were not compelled to narrow their focus to the extent that later organizations would be. In the more crowded political space that has existed in the United States since the 1960s, organizations are confronted with competing demands: the coalition contradiction discussed in chapter 5. Organizations initially seek to achieve the goals that they were formed to pursue. Because of the large number of organizations attempting to influence and win the attention of lawmakers, individual groups are now less capable of commanding attention and exercising influence on their own; thus the need for them to form coalitions. Movement organizations are obliged to seek common ground with others who share similar concerns in order to build the critical mass necessary to vie for the attention of lawmakers and to advance their goals. To make finding such common ground more likely, they may adapt their organizational goals to bring them more in line with those of other organizations. But the more that movement organizations have in common with one another, the more incapable they are of distinguishing themselves in their quest for funding and the popular support necessary to maintain the organization. Given the competing appeals for political and financial support directed towards the public and to funders by numerous organizations, soliciting organizations must offer a compelling reason why they, among many, should be rewarded. This forces SMOs to identify ways to distinguish themselves, including a narrowing of their organizational range to specific issues. But coalition partners are scarce for those focused on narrow concerns; thus the coalition contradiction.

Some organizations, such as those with more reliable resource flows than others or those that have more exclusive rights to represent specific interests, are less affected by the coalition contradiction. As discussed in previous chapters, labor unions, especially labor federations, are less constrained in terms of the goals they can pursue, since they can count on a steady flow of funds and serve as the sole legal representatives of certain interests. But voluntary organizations, such as environmental SMOs, which must constantly seek financial and member support, are subject to the competing pressures embodied in the coalition contradiction. Thus although the political and economic structures create the macro-structural framework within which social-movement groups operate, the different constraints that organizations experience within that framework require that we also focus on the organizational level of analysis to more fully comprehend the constraints on coalition activity. In the face of competing pressures to develop common ground with and distinguish themselves from others, organizations adopt various strategies, with some opting for a broader range and more coalition capability and others retaining a narrow focus, giving them a distinct identity but forcing them to forgo many opportunities for coalition participation.

Why some organizations select the narrow-focus option whereas others adopt broad goals and how they contend with the problems associated with their chosen paths are questions that require further research. Numerous factors influence the decision of each individual organization as to which of these two paths it should choose, but upon systematic examination, patterns among organizations would likely appear. In any case, organizations clearly do vary in terms of their range, and the evidence from this study suggests that those with a broader range of goals are more likely to engage in coalition activity.

Yet despite the assumptions of most coalition theorists, the goals pursued by SMOs, as represented by their organizational range, are not static. Rather, organizational learning can result in an expansion of organizational range. As described in chapter 6, organizational learning was found in this study to occur in one of two ways. The first form of organizational learning, experiential learning, occurs as a result of organizational failure (see figure 9.2). When standard strategies fail to produce the desired results, organizations experiment with other approaches. At times this experimentation may make movement participants aware of

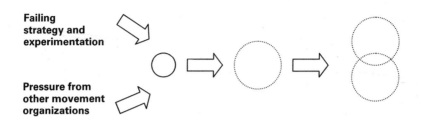

Failing strategy and experimentation

Pressure from other movement organizations

(a)
A failure of standard strategies in achieving organizational goals can lead to experimentation. On occasion this experimentation can lead organizations to adopt new goals, a form of organizational learning referred to here as "experiential learning." In addition, through competition or other forms of pressure, movement organizations can induce others to expand their organizational range.

(b)
Range expansion on the part of movement organizations creates more overlap with others and a greater potential for coalition formation. An overlap in range does not, however, necessitate that a coalition will form. Other conditions are necessary before coalition work can actually take place.

Figure 9.2

other issues to be addressed, which are then incorporated into the organization's established agenda.

An environmental organization working to protect tropical rain forests might provide an example of experiential learning. In a trial-and-error process of devising an effective strategy, the organization might encounter issues relating to the indigenous people who inhabit the rain forest, whose cultures are being disrupted by rain forest destruction. The resulting question of indigenous rights might start off as a proxy issue exploited by environmentalists in an effort to build support for rain forest protection, but it might also be recognized later as a legitimate issue that should be incorporated into the organization's routine work. If that happens, we can say that organizational learning has occurred leading to an expansion of the organization's range. This expansion, in turn, allows for more coalition possibilities, perhaps including work with human-rights organizations. In this study, some labor unions experi-

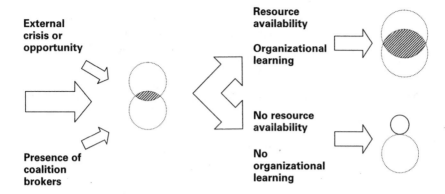

(c)
The presence of coalition brokers and external crises or opportunities create the conditions under which coalitions will actually form, usually among organizations that are similar in ideology and utilize the same tactics. The fill pattern indicates an overlap in range around which coalition activity actually occurs.

(d)
Organizational range can also be expanded through the learning that occurs in the process of interaction. If true organizational learning occurs, the range expansion will be made permanent, and the coalition is more likely to be institutionalized. The availability of resources to support coalition activity is another factor that can help to institutionalize a coalition. If learning only occurs among individual leaders, however, or if no resources are available to support coalition work, the participants are likely to return to routine practices, and coalition involvement will decline.

Figure 9.2
(continued)

menting with new strategies to confront abusive employers later adopted environmental issues when their innovative campaigns made them aware of environmental health risks to workers and the surrounding community.

Organizational learning can also occur as a result of the influence of other organizations. Social unions and environmental-justice groups have created pressure on the more narrowly focused organizations in their sectors to expand their range (Dreiling 1998). The mere presence of organizations utilizing a more encompassing approach or critiques by those organizations of their narrowly focused counterparts may cause those counterparts to change. Many "mainstream" environmental organizations responded to the critique launched by the environmental-justice movement by expanding their range to include more in the way of issues regarding the distributive element of environmental hazards. There may also be a competitive element at work here as organizations transform themselves to prevent the loss of members or support to those groups who are critical of their narrow agenda.

The expansion of an organization's range may result in greater overlap with that of other organizations, but this overlap in itself will not necessarily yield coalition activity. Although coalitions are numerous, not every organization that is working on the same issue (or very similar ones) is in coalition with every other. Those found to be most likely to work together are those that are similar in ideology and tactics. Although some have proposed that cooperating organizations will be tactically complementary—that is, they will use different tactics toward similar ends (Hathaway and Meyer 1997)—most empirical research suggests that organizations with similar tactics will more often engage in coalition activity together (Carroll and Ratner 1996; Hojnacki 1997; Staggenborg 1986). These tactical similarities create a compatibility that makes coordination between the organizations relatively unproblematic. Research has also indicated that coalition activity is more likely to occur when an external crisis or rare opportunity presents itself (Staggenborg 1986). In the context of labor-environmental cooperation, the efforts of various labor and environmental organizations in 1980 to build coalitions around the issue of preserving regulatory oversight by agencies like the EPA and OSHA can be seen as a reaction to the crisis of the election of Ronald Reagan, which presented a serious threat to the concerns of both movement sectors.

But there are still other conditions that must be met before coalitions form. Interorganizational contact does not occur spontaneously. In addition to the general pressures toward fragmentation associated with the U.S. political structure, legislative committees and administrative bodies usually focus on a specific policy area, reducing the frequency of contact between actors from different movement sectors. Although coalition work involving organizations within a single movement is relatively common due to the greater range overlap and routine contact among movement personnel, in order for organizations from different movement sectors to work together, individuals must actively coordinate the contact between them. As noted in chapter 7, these coalition brokers typically straddle the two movements in some way, enabling them to address both sides of a particular issue in ways that make sense. In the cases examined here, this might mean a labor leader with environmental sensibilities or an environmentalist with a union background. Third parties may also play the role of the coalition broker. This third-party brokering often involves organizations that address issues of concern to both sectors, such as citizens' groups or organizations that address issues of workplace health. Although intermittent contact between movement actors, each with a distinct focus, can occur, usually it is the presence of brokers that allows for coalition activity.

Intermovement contact is most likely to occur first around issues of overlapping concern to the two movements involved. Expansion of organizational ranges is significant in enabling this kind of initial contact to occur, because it creates more opportunities for organizations to find themselves at work on similar issues. The use of toxic substances in workplaces and the worker health and environmental risks posed by these substances served as the most common initial bridge issue for unions and environmental SMOs in this study.

Even when organizations are brought together, the trajectory of development of their relationship is contingent on many unpredictable microlevel processes. Despite all structural or organizational pressures to cooperate, individual incompatibilities among movement leaders may render cooperation impossible. Although such random factors can never be fully accounted for, some suggest that certain patterns can be anticipated. This is the argument made by cultural theorists, who see basic differences in the orientations of individuals from sectors as diverse as organized labor and the environmental movement (Rose 2000). Although some have offered evidence for these basic differences among rank-and-file participants in these two movement sectors, coalitions are in actual practice built not by rank-and-file members, but by organizational leaders and highly involved activists. Although it is possible to identify certain cultural tendencies among broad segments of the population, there is nevertheless variation among members within any group. Leaders and others who are very politically involved in their organizations can be distinguished from the majority of inactive members in those organizations. Among the characteristics that distinguish them are the very political sensibilities that enable them to see beyond narrow culturally proscribed views and to recognize their commonality with others working on related issues. Thus although fundamental cultural differences may exist between some rank-and-file unionists and the grassroots members of certain environmental organizations, the evidence presented here indicates that there are few cultural differences at the leadership levels of organizations, where coalitions are actually built, and that those differences that do exist are not significant barriers to cooperative activity.

Once organizations do manage to work together, even on a limited number of issues, the experience of doing so can give rise to another form of organizational learning, that derived from intermovement contact. Routine interaction with another organization, especially in the context of concerted efforts to achieve mutual goals, can build trust and understanding, which affects the individuals involved in the coalition work. Such contact tends to homogenize differences as each coalition participant learns more about the issues that are important to the other participants. In the process, the goals of one organization are in some cases incorporated into those of another, thus expanding that organization's range and allowing for still more coalition activity. In several cases examined here, both labor and environmental organizations adopted issues advanced by their coalition partners after a period of sustained contact and mutual effort.

But the extent to which such organizational learning actually takes hold is variable. The learning that individual coalition participants undergo must in some way be institutionalized within the organization in order for us to say that organizational learning has truly occurred. There are several ways that such institutionalization of individual learning can be achieved. Concerted efforts can be made by organizational leaders and coalition participants to educate the general membership about the new goals and understandings that have resulted from the learning. Successive leaders, who play a large role in determining the day-to-day focus and practices of any movement organization, can also be educated in this same way by their organizational predecessors. New goals can also be institutionalized by making them part of the group's formal platform or by the organization's formally joining an established coalition. Formal coalition membership and continued active participation in a coalition is enhanced if there are additional resources available to support them. If coalition work is simply added to the existing operations of an organization with no corresponding increase in resources or personnel to sustain it, the organization runs the risk that time and resource constraints will result in a return to routine activity centered on traditional objectives. However, if organizational learning truly occurs and resources are made available to support the additional activities associated with ongoing coalition participation, then there is a high probability that interorganizational relationships will be maintained and that coalition action will be institutionalized.

Social Movement Coalitions: Some Unanswered Questions

Using organized labor and the environmental movement as a basis for my analysis, I have attempted to flesh out the range of factors that determine the kinds of relations that develop between organizations and when cooperation among organizations can be expected. The distinctive characteristics of the labor and environmental movements require that attention be paid to several factors specific to relations between the organizations in both movements. I have also attempted, however, to identify larger patterns of intermovement and interorganizational relations that can be applied more broadly. But further research on other movements is necessary to verify the generalizability of the conclusions drawn here. I will now pose several questions that remain about the conclusions drawn in this study and offer suggestions for future research.

I have argued, earlier in this chapter and elsewhere in the book, that it was the proliferation of SMOs beginning in the 1960s that gave rise to the exceptional growth of coalition activity as a social-movement strategy. Although both the increase in the number of SMOs since the 1960s and the growth in coalition activity since that time have been empirically established, the existence of a relationship between these two factors needs further verification. Is the number of SMOs still increasing?

Has there been a corresponding increase in coalition activity? Or is coalition activity simply spreading through networks independent of the number of organizations, as is the case with other movement strategies (McAdam 1983)?

The analysis I have presented also indicates that organizational range is a major determinant of coalition participation. Those organizations that adopt a broader range of issues as their organizational focus are more likely to overlap with others and are therefore more likely to engage in coalition activity. This was clearly the case with the labor and environmental organizations examined here. It would be worthwhile to verify that this pattern is consistent in other movement sectors, but a more interesting question concerns why different movement organizations adopt a broad or narrow range and how they contend with the dilemma posed by the coalition contradiction. This question actually presents two related questions that require further research. The first is, how do organizations that embrace a broad organizational range distinguish themselves enough from other organizations to attract the support adequate to maintain the organization? Second, how do organizations with a very narrow range ever garner enough support to achieve their goals?

In regard to the first question, I offer several preliminary hypotheses. One possibility is that organizations with broad ranges distinguish themselves from other organizations with similar goals simply through marketing. Advertisers have long known that products can be promoted through image and style regardless of the actual content of the product. Presumably the same is true of SMO membership appeals. Slick promotional material or aggressive marketing techniques may allow an organization to stand out from others even when its agenda does not. Another possibility is that organizations distinguish themselves from others in areas other than the actual goals they are striving to achieve. Some have suggested that tactical differentiation is commonly used by organizations seeking to distinguish themselves from others. Issue overlap among such organizations would tend to favor coalition activity, although the barriers to cooperation among tactically distinct groups are well established (Staggenborg 1986, 1988). A third possible explanation for the survival of organizations that are nondistinct from one another in terms of the issues they address is that having a nondistinct issue agenda is an option truly available only to those organizations that are historically well established. The Sierra Club may be a good example of this. With a history spanning more than one hundred years, the Sierra Club has the name recognition and established networks necessary to draw support despite an agenda that overlaps considerably with those of several other organizations. Newer organizations may be more pressured to adopt a distinctive program than those already established. Research on newly forming SMOs could verify this hypothesis. Resource networks may also come into play as a factor in the survival of new breakaway organizations where the split with the parent organization is based more on individual personalities than any true divergence of tactics or goals. Well-known leaders or those with close ties to funders may be able to garner support for a new organization despite the lack of characteristics distinguishing it from others already in existence. A fourth possibility is that some broad-range organizations get a great deal of financial support from a small number of contributors. This begs the question of why these contributors would elect to support an organization that is otherwise indistinguishable from others, but the specific reasons why a small number of people may behave in such a way are more easily explained than would be the motivations of large numbers of people making this same seemingly arbitrary decision. A final proposal is that given a wide enough range, organizations are capable of attracting enough supporters from those interested in certain portions of their agendas to sustain the organization. For example, if a segment of the population supports only one particular plank of a group, most people in that segment will probably not join and will instead contribute to an organization focused specifically on their pet issue. But at least some percentage of the total support pool for that issue would probably be swayed to contribute to the broader organization anyway. By covering a large enough number of issues and attracting a fragment of the total support for each one, an organization may be capable of maintaining itself. This is especially likely since, given their broad range and the coalition support that they can command, these organizations could boast of more successes, thus making them a more attractive voice for a particular cause than a narrow, but less effective, group. Each of these hypotheses could be tested to determine why such a high degree of organizational redundancy exists within social movements.

The second and closely related question that arises out of the coalition contradiction has to do with how organizations that maintain a narrow range ever succeed at achieving their goals. Some research suggests that such organizations are actually more effective than those that maintain a broader agenda (Gamson 1990).[1] However, this has been found to be true only of organizations acting prior to 1945, before the proliferation of SMOs in the 1960s made coalition formation a more important strategy. Narrowly focused organizations have the advantage of being better able to distinguish themselves from others because of their well-defined and focused agendas, but given their narrow range, they have relatively little overlap with others, limiting the coalition aid they can receive to advance their goals. There are several possible explanations for how they may nonetheless achieve success. The first is that their range is never so narrow as to preclude coalition partners entirely. Even environmental SMOs that are focused on protecting a single species or a particular piece of land may be able to engage the support of broader environmental organizations for their modest and limited goals. We should also keep in mind that even organizations with narrow agendas may be able to engage in instrumental cooperation with dissimilar groups that have an unrelated interest in the same policy. An example was cited in chapter 5 of an alliance between tugboat workers and an environmental group focused exclusively on protecting the Puget Sound. Although both have fairly narrow agendas, they did share a common interest in requiring more control over tankers moving through the waters of the sound. Such instrumental coalition activity may be enough to bring organizations with narrow agendas success in some cases, despite the very narrow appeal of the goal being sought. A second possibility is that organizations with a narrow focus may achieve success because of a lack of opposition to their goals. Because of the crowded political field, narrowly focused organizations with few allies will not wield significant power, but if there is little or no resistance to what they are trying to accomplish, even weak organizations may be capable of advancing their agendas. A third possibility is that the specific issue addressed by a particular group has very wide or particularly intense support. The National Rifle Association may be an example of this phenomenon. Despite the association's having a very narrow agenda, its key issue, the right to bear arms, is deeply felt by many Americans, thus eliciting for the association a level of support with a sufficient intensity to enable the organization to persist and prevail even without an agenda expansive enough to attract a broad array of coalition partners. There may be an advantage to intensely concentrating on a single goal, if support for that one goal is broad and diverse coalition partners are not needed for success. Of course all of this is assuming that such narrow organizations do have a reasonable rate of success. An updated version of Gamson's comprehensive examination of social movement strategy would go far to verify the importance of organizational range and the necessity of coalition building as a movement strategy.

Another key finding of this study is the importance of the role of coalition brokers. Although other studies have found this role to be important as well, more research into social networks and how they relate to coalition activity would verify the extent to which these coalition brokers are necessary for coalition formation. Such study would also indicate where cross-movement personnel are most likely to be found, thus indicating where coalitions are liable to form.

Finally, this study has indicated that organizational learning and the associated expansion of organizational range has allowed for more coalition building among unions and environmentalists. This finding should be verified through the examination of other movement sectors. In addition, further research is needed to determine more specifically the conditions under which organizational learning and the institutionalization of coalition activity can be expected. The availability of resources to support coalition work is one factor found to contribute to coalition sustainability, but other issues associated with the institutionalization of cooperation should be explored. When are movement leaders more likely to try to educate their members about issues of concern to coalition partners, and when will they simply exploit ad hoc opportunities to work together with those partners on shared concerns? Under what conditions will they formally adopt additional issues as part of organizational agendas, and when will they simply go along for the sake of coalition maintenance?

The study of social movements largely lacks research that takes organizational behavior as its focus, and research that looks specifically at interorganizational and intermovement relations is even more limited. The traditional emphasis in social-movement study on individual recruitment and macrostructural conditions is certainly important and in many ways related to the issues examined here, and I have attempted to integrate such considerations into this analysis. But the behavior of organizations themselves needs further examination. The question here is not simply how SMOs recruit members and how they are affected by macrostructural changes, but also how they interact with one another and the strategies they adopt to advance their issues, including the formation of alliances with others. The study presented in this book has aimed to provide a useful framework for further exploration of these issues.

A Final Word on Unions and the Environmental Movement

There is a clear need for more analysis of all social movements at the organizational level. I believe, however, that because of the great potential posed by the cooperation of the two movements involved, the study of labor-environmental relations is, in itself, worthy of further examination. Historically, organized labor has been the driving force behind much of the social change that has taken place since the industrial revolution. Although organized labor has been written off as an irrelevant relic by some and almost completely neglected by social movement scholars of the last three decades, it is likely that organized labor will continue to be an indispensable player in many of the major social transformations to come. The basic relations of production that characterize capitalist economies remain unchanged, and it is likely that the class antagonism that this fosters will continue to provide a role for organized labor as long as those relations remain in place.

But the social changes that have taken place especially since the 1960s have also opened a space for numerous other movements to develop around a vast array of interests and identities. Although some of these movements arouse the interest of small segments of the population, one that garners perhaps more support than any other is the environmental movement. The scope of the threat presented by environmental degradation may be responsible for the broad support that has developed for protection of the natural environment. With the increasingly global nature of some environmental problems, virtually no one can escape the dangers they pose. But although everyone is in the same global boat, some are still identifiable as the primary perpetrators of ecological ruin. Though there are many who benefit materially from the excessive use of resources that fosters environmental problems, the costs and benefits of such use are not distributed equally, nor is the ability to alter these use patterns. Those who control the means of production have a greater ability to change the way in which goods are produced, and perhaps not surprisingly, they disproportionately benefit from the environmentally destructive practices currently in use. Thus similar battle lines are drawn to those organized around employment. Although organized into different sectors within the fragmented political framework that is U.S. democracy, both unions and environmental SMOs are aligned against the same forces. When divided, they represent relatively weak movements in relation to the power wielded by those private entities engaged in the exploitation of workers and nature. Yet when allied, unions and environmentalists represent a force capable of offering a significant counterweight to their mutual adversaries. An alliance between these two sectors on a comprehensive and lasting basis would represent a movement capable of bringing about dramatic social change. Thus, social-movement scholars and strategists should give serious consideration to the relations between these movement sectors and the important social transformation they could potentially bring.

Appendix: Research Methodology

The research I have presented in this book was carried out in two phases. The first phase consisted of a survey of state labor federation leaders. This survey was used to gauge the quality of relations between unions and environmentalists in each state and to identify key issues around which relations between these two movements developed. The survey data were also analyzed to assess the association between political and economic conditions and the quality of labor-environmental relations. The second phase of this research project involved qualitative analysis in which states were selected for case study. Interviews with over 70 labor, environmental, and other social movement leaders and activists along with organizational documents were used as the basis for data analysis. Here the intent was to understand the process of intermovement relationship development.

The first phase of this study was published in *Social Science Quarterly* in March 2002. I cite this research several times in this book, but the details of this portion of the research are not included, since they are available elsewhere. I will provide a brief description of that methodology here, however. More details are available in the *Social Science Quarterly* article.

The survey of state labor federation leaders was conducted between December 1997 and July 1998. Responses were obtained from forty-seven of the fifty state labor federations either through a mailed questionnaire or follow-up telephone calls. State labor leaders were selected as the target of the survey because of the relatively centralized structure of the labor movement. Whereas a number of different environmental groups may populate a particular state, the state labor federation serves as the umbrella organization for all affiliated unions. Given their role as the primary political advocates for the labor movement in a state, any significant labor-environmental conflict or cooperation in the political arena would involve state federation leaders, making them well positioned to offer an assessment of the overall quality of labor-environmental relations in the state. Although a corresponding assessment of the labor-environmental relationship from the perspective of environmental leaders would be useful, because of the presence of multiple environmental organizations in each state and the lack of any centralized structure in the environmental movement, no consistent and systematic means were available for obtaining such an assessment. Given this limitation and given the centralized position of state labor leaders, the assessment offered by the labor leaders was taken as the best and most consistent measure available for assessing the condition of labor-environmental relations in each state. Additional interviews with some state environmental leaders conducted in 1998–1999 were used to confirm the status of labor-environmental relations, and the outcomes of these interviews were consistent with the characterizations offered by labor leaders.

For the survey portion of the study, those union leaders responsible for the political work of the federation (legislative directors, lobbyists, officers, etc., depending on the particular federation) in each state were asked about the relationship between their federation and environmental organizations and whether they had engaged in such activities as information sharing, holding regular meetings, issuing joint press statements, coordinated lobbying, or other forms of joint political action with those environmental organizations. They were also asked about specific environmental issues in which they were involved and about their relations with employers regarding environmental issues. The dependent variable in the statistical analysis was defined by the union representative's characterization, on a five-point scale ranging from "very good" to "very bad," of the quality of the relations between their federation and environmentalists.[1] This information, reported in chapter 2, was used as a basis for exploring issues in the qualitative phase of the project.

Beyond the generation of descriptive data, the survey data were also used as the basis for a regression analysis. State level economic and political indicators were used to identify factors associated with the quality of the labor-environmental relations in each state. The percentages of workers employed in the construction, timber, primary metal, paper, chemical, and service sectors were used as independent variables in the model. These employment sectors were identified as being among those most likely to be affected, either positively or negatively, by environmental policy; thus the larger the presence of workers in these sectors in a state, the more likely relations between labor and the environmental community generally will be influenced. Based on the way environmental policy can affect these employment sectors, it is reasonable to assume that the percentage of a state's workforce employed in a given industry will affect the actions taken by that state's labor federation in regard to environmental issues.[2] The U.S. Census Bureau's County Business Pattern data (U.S. Bureau of the Census 1997) were used to determine the employment percentages in the relevant industries.

In addition to the employment sectors, variables associated with the political conditions in the states were also included in the model. A measure of partisan control at the state level was derived by assigning one point each for Republican control of the lower house, the upper house, and the governor's office. Thus, a four-point scale (0–3) of partisan control was created for each state. A score of zero represented a state in which Democrats held a majority in both houses and had control of the governor's office. A score of three meant that Republicans controlled both houses and the governor's office. The measure was based on partisan control in 1996 as reported in the *Statistical Abstracts of the United States, 1997* (U.S. Bureau of the Census 1998).

In addition to this simple measure of partisan control, a variable was also included to represent the type of party system in each state. This refers to the base of support that underlies the state parties and the associated social divisions within the state (Brown 1995). For example, although unions serve as a base of support for the Democratic Party in virtually every state, their significance within state parties varies considerably from state to state. Union density was added to the model as a way to account for union strength in each state, but an additional variable related to the party system was also included. Differences in the party system in terms of the base of support for the state political parties give rise to distinct partisan cleavages. In one, the "southern partisan cleavage," race plays the most significant role, whereas in others, such as the "New Deal party cleavage," traditional class divisions are more prominent. These differences in party systems could influence the way in which organizations within the state relate to one another.

A related variable is that of ideology. Although ideology can be a particularly nebulous factor and one that is difficult to gauge, some scholars have generated a measure of citizen and government ideology that can allow for some control of ideological influences at the state level (e.g., Berry et al. 1998). A measure of citizen ideology was also included in the quantitative analysis.

The final variable included was one based on the triangular political configuration generated by the presence of environmental advocates, polluters, and the unionized employees of polluting firms. Given that unions are thought to occupy a middle ground between the primary adversaries, environmentalists and environmentally destructive industries, labor-industry relations may influence the type of relations that exist between workers and environmentalists. A measure of labor-industry cooperation was included as a means of testing the notion that unions are in a position to "choose sides" in relation to environmental issues. This measure was based on a survey question in which union leaders were asked whether their state federation had engaged in cooperative efforts with industry organizations on environmental issues.

The results of the data analysis indicated that three of the independent variables have a significant effect on labor-environmental relations. The model shows that the percentage of workers employed in the timber industry, partisan control of the state government, and labor's record of having worked with industry on environmental issues all have a significant effect on the quality of labor-environmental relations in that state. These results, along with the relevant findings of other researchers, were all incorporated into the discussion and analyses in chapters 2 and 3. No significant association was found between the percentage of workers employed in the paper, chemical, primary metal, service, or construction sectors and labor-environmental relations. Although partisan control of a state's government is associated with the quality of the state's labor-environmental relations, party cleavages were found to have no significant effect on those relations. Union strength and citizen ideology were also found to have no association with labor-environmental ties.

The survey data and the corresponding quantitative analysis provided important insight into the overall quality of labor-environmental relations from state to state while also demonstrating the limitations of a simple interest framework. Basic political and economic interests account for some of the variation in the quality of labor-environmental relations, but much is left unexplained by these variables. The second phase of the study, which constitutes much of the material presented in this book, was designed to help identify the complex dynamics that lie behind the statistical patterns and to show how particular processes can yield outcomes that differ from the patterns identified. Factors investigated in the case studies conducted in the research's second phase include the level and quality of interaction between labor and environmental leaders, the effect of the organizational structure of particular labor and environmental organizations on interorganizational relations, and the role played by the leadership and core activists within these movement sectors.

As with the survey, states served as the initial unit of analysis for the case studies, which Yin (1994) calls a "holistic" case study design. Five states were examined in depth, with the national level serving as an additional case. Although some inferences are drawn from the states as a whole, most of the analysis focuses on relations between particular environmental and labor organizations within each state in what is referred to as "embedded case" analysis (Yin 1994). Because of the variation that exists among movement organizations even within the context of a single state, it is most important to examine factors at the organizational and interorganizational levels of analysis.

The multiple-case design employed here enables the detailed examination of several factors thought to shape the quality of intermovement relations. Four factors were expected to affect interorganizational relations in each state, yielding four general hypotheses that oriented the research. The first factor is the breadth of focus of the movement organizations themselves. Some organizations focus their agenda on a narrow set of issues relevant to their members, whereas others have a more encompassing vision in which they address a broader range of issues and attempt to advance the interests not just of their members, but of some larger segment of society or the society as a whole. It was hypothesized that organizations with broader agendas would have better relations with one another. The second factor concerns the amount and type of organizational interaction. It was predicted that greater organizational interaction would expand the agenda of participating organizations and improve relations. More interaction, especially of a routine and structured kind, was predicted to improve relations and to "homogenize" organizational agendas. The third factor, the presence of coalition brokers, or actors who in some way incorporate the values and goals of both movements, was hypothesized to have a positive effect on the quality of relations between unions and environmentalists. Those who are in some way grounded in both movements can play the role of mediating differences and of actually bringing the two movement sectors into contact. The fourth factor scrutinized is the cultural differences that exist between the two sets of actors. Other scholars have identified the way in which class-cultural differences can serve to divide unionists and environmental activists. These differences, however, have primarily been identified among grassroots participants in movement organizations. I hypothesized that cultural barriers between organizational leaders would not be as significant as those among grassroots participants, posing few obstacles for intermovement cooperation at the state level.

In addition to these explanatory efforts, there are also *exploratory* elements to the approach used here. Throughout the research process I remained cognizant of other, unexpected influences and incorporated them into the research as they became evident.

The five states used as the basis for case study include Maine, New Jersey, New York, Washington, and Wisconsin. Some investigation of labor-environmental relations was also conducted at the national level, in part to identify whether and how national activity influenced relations at the state level, but also to identify any differences in interorganizational dynamics at this level of organization. States were selected based on certain similarities that some had with one another and on important differences that would allow for comparison. Cases come from the Pacific Northwest, the Midwest, and the Northeast, providing some regional variation. The characterization by union leaders in the five states of the labor-environmental relationship based on the initial survey, range from "good" to "poor". In terms of the partisan control of the state government, at the time of the study the states ranged from almost complete Democratic Party control (Maine) to complete control by Republicans (New Jersey). Three of the states included in the study have above-average levels of timber industry employment, with Washington having almost three times the national average; the remaining two (New York and New Jersey) have a negligible percentage of timber industry employment. Union leaders from each state except Wisconsin reported having worked with industry on environmental measures. The presence of this kind of variation among the states included in the study allowed for consideration of the influence of each variable when appropriate.

Consideration was also given to states known to have labor-environmental relations of special interest that could possibly shed light on the process of relationship formation. Wisconsin is known to have a long history of labor-environmental cooperation and an existing (although now dormant) coalition of unions and environmental organizations. Washington was greatly affected by the timber industry disputes between environmentalists and loggers in the 1980s and 1990s, which the survey suggested had a significant effect on labor-environmental relations. Yet the labor movement in Washington is also known for being particularly progressive and has a history of outreach and coalition formation. Maine also faced an extremely controversial series of measures regarding forest policy during the 1990s, yet labor unions and most of the state's environmental organizations were able to maintain positive ties throughout that period. As with all research, cases were also selected based on convenience and the limitations imposed by the financial resources available.

In conducting the case studies, preliminary research was done through the use of secondary sources that examined labor and environmental issues in each state. This provided background and the names of some organizations and individuals to contact. Identifying relevant labor leaders in the five states was made easier because of the centralized nature of the labor movement at the state level. Labor representatives associated with the political work of the state federation served as the initial contact for the labor side. Using a snowball sampling method, these informants were also used as a basis for identifying other relevant figures in the labor and environmental communities.

Various web search engines were used to identify environmental organizations in the five states selected. Where organizations had their own web pages, these sources were used to gather information on the relevant environmental issues in the state, to identify factors suggesting the organization's relations with the labor community, and to identify individuals and contact information. A list of initial contacts was generated from these sources, and the individuals on that list were contacted first by mail, then by phone, to request an interview. During the initial phone conversation they were also queried as to other relevant individuals to contact, and the procedure was repeated. Based on the recurrence of individual names and organizations, I was able to identify the most politically active organizations in each state and at least some of the key players in labor-environmental politics. Statewide organizational leaders were targeted because of their important role in the development of state environmental and economic policy. Where local environmental organizations or labor representatives were identified as having played a significant role in labor-environmental relations in the state, however, they were sought as well.

Where it was possible to arrange a meeting with labor or environmental representatives, interviews were conducted, focusing on several aspects of the group that they represent and their relations with other organizations. Questions centered on four main areas that correspond to the hypotheses being tested. In regard to the issue of the breadth or narrowness of the organizational agenda, informants were asked what issues they worked on, including whether or not labor/environmental issues ever arose in their work. To provide further insight into this issue, questions about the organizational features of each group were also posed (the structure of the organization, decision making procedures, policies, etc.). The second area of questioning focused on the amount and type of interaction with movement organizations from the other sector. The third issue, that of the role of coalition brokers, included a set of questions regarding how interorganizational contact was initially established and who carried out that contact. Lastly, participants were asked about any stylistic differences they perceived in the operation of the organizations in the other movement (i.e., labor leaders were asked about perceived stylistic differences in environmental organizations, and environmental leaders were asked about perceived stylistic differences in labor unions).

Interviews were conducted in a way that is best described as a combination of open-ended and focused interview methods. There was a specific set of questions for all interviewees, but there were also portions in which broad questions explored both matters of fact and the opinion of the subjects. Thus there were at times in the interviews elements of focused and formal interviewing, and at other times the questions were open ended and the manner conversational.

Analysis of the interview data began with a thorough review of the interview transcripts. During this initial review I identified themes derived from the original hypotheses and searched for other relevant issues to explore. Although no new hypotheses were developed as a result of transcript review, several recurring themes were noted for further consideration, including job blackmail (threats of job loss used by employers to pressure unions to oppose environmental measures), electoral issues, organizational limitations (practical resource or time constraints faced by movement organizations), and "greenwashing" (efforts by polluters to portray themselves as environmentally responsible). In addition, I did expand and refine the hypotheses during this stage of the study. For example, in regard to the hypothesis concerning the transformation of organizational goals as a result of organizational interaction, I respecified this transformation as "organizational learning" and broke the process down into "learning through interaction" and "experiential learning." The data offered evidence of goal and strategy transformation that resulted from trial-and-error experiences that organizations underwent unilaterally, thus the addition of experiential learning as a distinct category of learning to be explored.

In the next phase of the data analysis, the Non-numerical Unstructured Data Indexing Searching and Theorizing (NUD*IST) software package was used to code and index the interview data. Codes were created for each of the central themes identified during transcript review, and interview segments were indexed accordingly. These indexed segments were then reviewed to identify patterns among the views expressed and the information provided and the organizations to which the interviewees belonged. As with all data, both qualitative and quantitative, some inconsistencies were found. I primarily report on the patterns that clearly emerged; where appropriate, however, I include discussion of contradictory evidence.

It is widely acknowledged that a variety of methodological approaches can increase the validity of research findings. Such "triangulation" was achieved in this study in several ways. First, the case studies themselves were used to supplement the quantitative analysis conducted in the first phase of the study. The subjective assessments regarding the quality of labor-environmental relations and the descriptions of issues and experiences with other organizations provided by the initial survey informants were confirmed through interviews with leaders of both organizational sectors in the second phase of the study. But in addition to the two-stage design of the overall project, several sources were used within the case studies themselves. Interviews served as the primary source of information, and interviews with a variety of subjects in regard to the same issues provided a means to validate reports. Organizational literature and archives were also used to verify certain data provided by interviewees and to provide supplemental information about issues and organizational operations. In some cases access to a comprehensive organizational archive was obtained, complete with documents internal to the organization and personal communications. In other instances only general organizational literature (and information posted on the web) provided for the general public was obtained. Systematic observation was not a part of the research design; I did, however, sit in on one coalition meeting in Maine, and in some instances I interviewed more than one subject simultaneously. This enabled me to observe the interaction among certain actors and to hear how they discussed the issues with one another throughout the course of the interview.

Notes

Chapter 5

1. The study presented in this book focuses exclusively on the United States. Under different structural conditions than those present in the United States, we should expect very different political dynamics among social movement organizations and political parties. In addition, the historical conditions under which movements developed in other nations has also given some a different character relative to those in the United States. The most obvious example is that of the labor movement. Scholars have noted that unions formed under conditions different from those in the United States often have a stronger "social union" orientation (Dreiling and Robinson 1998; Seidman 1994).

2. As will be discussed later, unions are unique among social movement organizations because they do have an officially recognized role in the workplace.

3. The AFL-CIO's declining influence over the Democratic Party must be understood as not simply a decline in relative influence due to the growth of other movements that serve as a base of support for the party. Labor has also lost power in absolute terms because of the greatly diminished membership in unions.

4. In this progression, higher forms have not displaced the lower forms of representation but instead indicate the growing need for broader and more coordinated forms of representation in addition to the narrower and more individually focused forms of the past.

5. There is a great deal of debate about the entrepreneur analogy and the nature of professional movement organizations. Some organization theorists argue that organizational actors, even in nonprofit enterprises, readily lose sight of the organization's original goals and instead become focused on maintaining the organization and the benefits they accrue from their positions (Michels 1949; Rucht 1999; Scott 1992). This is certainly the implication of the analogy to business entrepreneurs. Though useful to consider, however, the business analogy and the emphasis placed upon organizational maintenance for its own sake has its limitations when applied to movement organizations. Although profit is typically the central motivation behind a business enterprise, political entrepreneurs are

motivated largely by a commitment to their cause (Sabatier and McLaughlin 1990). Political entrepreneurs typically participate in movement activity for a long period of time without compensation before they attain professional positions within an organization. And even though movement professionals may displace the original entrepreneur (Staggenborg 1988), their ideological commitment has been demonstrated by attitudinal research and reinforced by the fact that professional movement actors are compensated at a far lower rate than those doing similar work in other fields (Berry 1978; Schlozman and Tierney 1986). In addition, effort expended to maintain an organization does not in itself necessarily indicate a shift in priorities away from the organization's original political objectives, since the organization is viewed as instrumental in pursuing those objectives. The preservation of an organization cannot be separated from the pursuit of the goal for which the organization was created.

6. The organizational structure of a group may determine where the decision regarding whether to undertake coalition work and accept the compromises it entails will be made. Because the focus of the study presented in this book is on statewide organizations and state-based affiliates of national groups, many of the environmental organizations considered have similar formal decision-making processes and an organizational mission statement that defines the basic purpose of the group. Within the framework provided by the mission statement and decision processes, a statewide board will be formally in charge of setting the organization's policy. The board may be elected directly by the membership, it may include representatives of local chapters (where they exist), or it may simply be a collection of concerned and involved people recruited by the organization's founder. The role of the board is to provide general oversight of the organization while the executives carry out the day-to-day decision making and administrative responsibilities.

Although the organizational structures of environmental groups may vary, decision making regarding the specific strategies an organization should use and the policies it should support often fall to a relatively small group of very involved people in the organization, a tendency that is long recognized by organizational theorists (Michels 1949; Scott 1992). Most of the people interviewed for this study were full-time staff members. Although some express a desire that their organization not be a "staff-run" organization, the fact is that much of the day-to-day decision making is carried out by these individuals, with more or less input from a formal board.

Organizations vary in the extent to which their boards are actively involved in decision making or simply rubber-stamp the actions of the organizations' executives and staff, yet regardless of the point of decision making, the conclusions described by most of the environmental advocates interviewed was strikingly similar. In terms of collaborative work with labor unions, most describe themselves as being sympathetic to the union cause and willing to work with labor when possible but acknowledge that organizational constraints, to varying degrees, prevent them from veering too far from their organizations' stated missions.

7. International union organization based in a particular industry no longer applies in many cases. Some unions strive to organize workers in areas very different from their original base. For example, teaching assistants at the University of California are affiliated with the United Auto Workers. Most "international" unions are also based exclusively in the United States.

8. In some states workers may still be required, even if they do not join the union, to pay dues to the union to cover the cost of the services that the union provides (collective bargaining, grievance procedures, etc.); the decision not to formally become a union member limits these workers' access to participation in union activities.

9. This focus is to some extent mandated by law. However, such a mandate applies only to collective bargaining. In terms of their political work, unions are free to pursue other issues, but many still remain focused on narrow concerns of members. This narrow focus of unions will be taken up in greater detail in a later chapter.

Chapter 6

1. The iron law of oligarchy may be more applicable to labor unions, for reasons described in chapter 5. Many union members did not actually seek out membership in their union but simply signed up given the existing presence of the union at their workplace. Thus membership in a union is often of a different quality then membership in a purely voluntary social movement organization. This can make labor leaders less susceptible to grassroots pressure, as evidenced by the tendency for existing leaders to retain their position of power for decades. Leadership of other social movement organizations tends to change more regularly.

Chapter 7

1. Becuase this research subject requested anonymity, this is not his real name.

Chapter 8

1. Of course, the shift to new issues can occur only within the specified organizational range as discussed in chapter 5. An organization dedicated to ocean protection, for example, cannot readily jump to rain forest issues simply because they are salient at a given moment.

Chapter 9

1. Gamson also suggests that this finding may be due to the fact that in his sample the multiple-issue organizations had a more extreme goals: displacing

their antagonists. Once this factor is controlled for, he finds no independent effect of multiple-issue demands on success rates.

Appendix

1. Although the environmental movement is composed of numerous organizations that may vary in their positions and relations with labor unions, the union representatives were asked generally about relations with environmentalists in their state. None expressed difficulty in offering such an assessment or offered a mixed response in regard to different environmental organizations; thus their responses are considered to be an accurate measure of the overall quality of labor-environmental relations in their state. Union representatives were later asked to identify particular organizations and to discuss their relations with them, at which time a few did identify different organizations and some relations inconsistent with the overall pattern of relations they had identified.

2. Although measures of *unionized* employees by sector would be a more accurate measure for this variable, state federations often serve as the voice of worker interests even for nonunionized workers or unaffiliated unions. In any case, state level data on union density by sector was not readily available.

References

Ackerman, Bruce, and William Hassler. 1981. *Clean coal, dirty air*. New Haven: Yale University Press.

AFL-CIO Executive Council. 1997. Portland, Oregon, February 20. ⟨www.aflcio.org/aboutaflcio/ecouncil/statements.cfm⟩.

Alliance for Sustainable Jobs and the Environment. 1999. "Houston Principles." Millennium Edition. Vol. 1, no. 1.

Alliance for Sustainable Jobs and the Environment. 2000. Vol. 2, no 1.

Almond, Gabriel, and Sidney Verba. 1972. Organizational membership and civic competence. In *Group Politics*, ed. Edward Malecki and H. R. Mahood, 25–44. New York: Scribner's.

Altemose, J. Rick, and Dawn A. McCarty. 2001. Organizing for democracy through faith-based institutions: The Industrial Areas Foundation in action. In *Forging Radical Alliances across Difference*, ed. Jill Bystydzienski and Steven Schacht, 133–145. London: Rowman and Littlefield.

Alvarez, Lizette, and Joseph Kahn. 2001. House Republicans gather support for Alaskan drilling. *New York Times*, August 1, p. A13.

Argyris, Chris, and Donald Schön. 1978. *Organizational learning*. Reading, MA: Addison-Wesley.

Arrandale, Tom. 1994. The Sagebrush Gang rides again. *Governing* 7(6): 38–42.

Associated Press. 2002. Automakers, union, plan meetings to protest new fuel economy standards. February 23.

Audley, John. 1995. Environmental interests in trade policy: Institutional reform and the North American Free Trade Agreement. *Social Science Journal* 32(4): 327–360.

Axelrod, Robert. 1986. Presidential election coalitions in 1984. *American Political Science Review* 80: 281–285.

Bacharach, Samuel, and Edward Lawler. 1980. *Power and politics in organizations*. San Francisco: Jossey-Bass.

Bailey, Eric. 1999. Loggers try to carve a new niche. *Los Angeles Times*, December 1, p. A3.

Banks, Andrew. 1990. Jobs with justice: Florida's fight against worker abuse. In *Building bridges: The emerging grassroots coalition of labor and community*, ed. Jeremy Brecher and Tim Costello, 25–37. New York: Monthly Review.

Baumol, William, and Wallace Oates. 1979. *Economics, environmental policy and the quality of life*. Englewood Cliffs, NJ: Prentice Hall.

Bell, Sandra J., and Mary E. Delaney. 2001. Collaborating across difference: From theory and rhetoric to the hard reality of building coalitions. In *Forging radical alliances across difference*, ed. Jill Bystydzienski and Steven Schacht, 63–76. London: Rowman and Littlefield.

Berry, B. J. 1977. *The social burdens of environmental pollution*. Cambridge, MA: Ballinger.

Berry, Jeffrey, 1978. "On the Origins of Public Interest Groups." *Journal of Politics* 10:392–393.

Berry, Jeffrey. 1984. *The interest group society*. Boston: Little, Brown.

Berry, William D., Evan Rinquist, Richard Fording, and Russell Hanson. 1998. Measuring citizen and government ideology in the American states, 1960–93. *American Journal of Political Science* 42(1): 327–348.

Bevacqua, Maria. 2001. Anti-rape coalitions: Radical, liberal, black and white feminists challenging boundaries. In *Forging radical alliances across difference*, ed. Jill Bystydzienski and Steven Schacht, 163–176. London: Rowman and Littlefield.

Bobo, Kim, Jackie Kendall, and Steve Max. 2001. *Organizing for social change*. 3rd ed. Santa Ana, CA: Seven Locks.

Bourdieu, Pierre. 1990. *Outline of a theory of practice*. New York: Cambridge University Press.

Braverman, Harry. 1974. *Labor and monopoly capital*. New York: Monthly Review.

Brecher, Jeremy, and Tim Costello. 1990. *Building bridges: The emerging grassroots coalition of labor and community*. New York: Monthly Review.

Brecher, Jeremy, and Tim Costello. 1994. *Global village or global pillage*. Cambridge, MA: South End.

Broder, David. 1978. Introduction. In *Emerging coalitions in American politics*, ed. Seymore Martin Lipset, 3–11. San Francisco: Institute for Contemporary Studies.

Brody, Evelyn. 1996. Agents without principals: The economic convergence of the non-profit and for profit organizational forms. *New York Law School Law Review* 40: 457–536.

Bronfenbrenner, Kate, Sheldon Friedman, Richard Hurd, Rudolph A. Oswald, and Ronald L. Seeber. 1998. *Organizing to win*. Ithaca, NY: ILR.

Bronfenbrenner, Urie. 1972. Socialization and Social Class. In *The Impact of Social Class*, ed. Paul Blumberg, 381–409. New York: Crowell.

Brown, Beverly. 1995. *In timber country.* Philadelphia: Temple University Press.

Brown, Robert. 1995. Party cleavages and welfare effort in the American states. *American Political Science Review* 89(1): 23–33.

Brulle, Robert. 1996. Environmental discourse and social movement organizations. *Sociological Inquiry* 66(1): 58–83.

Bryant, Bunyon. 1995. *Environmental justice.* Washington, DC: Island.

Bryner, Gary. 1995. *Blue skies, green politics.* Washington, DC: CQ Press.

Buchanan, Millie, and Gerry Scoppettuolo. 1997. Environment for cooperation: Building worker-community coalitions. In *Work, Health and Environment,* ed. Charles Levenstein and John Wooding, 329–351. New York: Guilford.

Bucholtz, Barbara. 1998. Reflections on the role of non-profit associations in a representative democracy. *Cornell Journal of Law and Public Policy* 7: 555–603.

Bullard, Robert. 1993. Anatomy of environmental racism and the environmental justice movement. In *Confronting Environmental Racism,* ed. Robert Bullard, 15–39. Boston: South End.

Burton, Dudley. 1986. Contradictions and changes in labour responses to distributional implications of environmental-resource policies. In *Distributional Conflicts in Environmental-Resource Policy,* ed. Allan Schnaiberg, Nicholas Watts and Klaus Zimmerman, 287–314. Berlin: Gower.

Buttel, Frederick. 1975. The environmental movement: Consensus, conflict and change. *Journal of Environmental Education* 7: 53–63.

Buttel, Frederick. 1986. Economic stagnation, scarcity, and changing commitments to distributional policies in environmental resource issues. In *Distributional conflicts in environmental-resource policy,* ed. Allan Schnaiberg, Nicholas Watts, and Klaus Zimmerman, 221–238. Berlin: Gower.

Buttel, Frederick, Charles Geisler, and Irving Wiswall. 1984. *Labor and the environment.* Westport, CT: Greenwood.

Buttel, Frederick, and Oscar Larson. 1980. Whither environmentalism? The future political path of the environmental movement. *Natural Resources Journal* 20: 323–344.

Bystydzienski, Jill, and Steven Schacht, eds. 2001. *Forging radical alliances across difference.* London: Rowman and Littlefield.

Carlton, Jim. 1999. Unions, environmentalists unite to pressure for jobs, resources. *Wall Street Journal,* October 4.

Carroll, Mathew, and Robert G. Lee. 1990. Occupational community and identity among Pacific Northwestern loggers: Implications for adapting to economic changes. In *Community and forestry: Continuities in the sociology of natural resources,* ed. Robert G. Lee, Donald Field and William Burch, 141–155. Boulder, CO: Westview.

Carroll, W. K. and R. S. Ratner. 1996. Master framing and cross-movement networking in contemporary social movements. *Sociological Quarterly* 37(4): 601–625.

Carson, Rachel. 1962. *Silent Spring*. Cambridge, MA: Riverside.

Chertkoff, Jerome. 1970. Sociopsychological theories and research on coalition formation. In *The study of coalition behavior*, ed. Sven Groennings, E. W. Kelley, and Michael Leiserson, 297–322. New York: Holt, Rinehart and Winston.

Chertkoff, Jerome. 1976. Sociopsychological views on sequential effects in coalition formation. In *Coalitions and Time*, ed. Barbara Hinckley. Beverly Hills, CA: Sage.

Childs, John Brown. 1990. Coalitions and the spirit of mutuality. In *Building bridges: The emerging grassroots coalition of labor and community*, ed. Jeremy Brecher and Tim Costello, 234–242. New York: Monthly Review.

Clark, Peter, and James Q. Wilson. 1961. Incentive systems: A theory of organizations. *Administrative Science Quarterly* 6: 129–166.

Clawson, Dan, and Mary Ann Clawson. 1999. What has happened to the US labor movement? *Annual Review of Sociology* 25: 95–119.

Cohen, Joshua, and Joel Rogers. 1983. *On democracy*. New York: Penguin.

Collie, Melissa P. 1984. Legislative voting behavior. *Legislative Studies Quarterly* 9: 3–50.

Connors, Tracy Daniel. 1997. *The non-profit handbook*. New York: Wiley.

Cooper, Mary. 1992. Jobs vs. the environment. *CQ Researcher* 2(18): 411–431.

Cotgrove, Stephen, and Andrew Duff. 1980. Environmentalism, middle class radicalism and politics. *Sociological Review* 18: 333–351.

Crotty, William J. 1984. *American parties in decline*, 2nd ed. Boston: Little, Brown.

Cullen, Pauline P. 2001. Coalitions working for social justice: Transnational NGOs and international governance. In *Forging radical alliances across difference*, ed. Jill Bystydzienski and Steven Schacht, 249–263. London: Rowman and Littlefield.

Davis, Charles, and Sandra Davis. 1988. Analyzing change in public lands policymaking: From subsystems to advocacy coalitions. *Policy Studies Journal* 17: 3–24.

Dery, David. 1998. "Papereality" and learning in bureaucratic organizations. *Administration and Society* 129(6): 677–689.

Dewey, Scott. 1998. Working for the environment: Organized labor and the origins of environmentalism in the United States: 1948–1970. *Environmental History* 1: 45–63.

Domhoff, G. William. 1967. *Who rules America?* Englewood Cliffs, NJ: Prentice Hall.

Dowie, M. 1995. *Losing ground: American environmentalism at the close of the twentieth century.* Cambridge: MIT Press.

Downey, Gary. 1986. Ideology and the clamshell identity: Organizational dilemmas in the anti–nuclear power movement. *Social Problems* 33: 357–373.

Downs, G. W., and L. B. Mohr. 1979. Toward a theory of innovation. *Administration and Society* 10(4): 379–408.

Dreiling, Michael. 1997. Remapping North American environmentalism: Contending visions and divergent practices in the fight over NAFTA. *Capitalism, Nature and Socialism* 8(4): 65–98.

Dreiling, Michael. 1998. From margin to center: Environmental justice and social unionism as sites for intermovement solidarity. *Race, Class and Gender* 6(1): 51–69.

Dreiling, Michael, and Ian Robinson. 1998. Union responses to NAFTA in the US and Canada: Exploring Intra- and international variation. *Mobilization: An International Journal* 3(2): 163–184.

Dunk, Thomas. 1994. Talking about trees: Environment and society in forest workers' culture. *Canadian Review of Anthropology and Sociology* 31(1): 14–34.

Echeverria, John, and Raymond Booth Eby, eds. 1995. *Let the people judge.* Washington, D.C.: Island.

Eder, Klaus. 1993. *The new politics of class.* London: Sage.

Epstein, Barbara. 1991. *Political protest and cultural revolution.* Berkeley and Los Angeles: University of California Press.

Fantasia, Rick. 1988. *Cultures of solidarity.* Berkeley and Los Angeles: University of California Press.

Ferree, Myra Marx, and Frederick Miller. 1985. Mobilization and meaning: Toward an integration of social psychological and resource perspectives on social movements. *Sociological Inquiry* 55(1): 38–61.

Field, Rodger. 1998. Risk and justice: Capitalist production and the environment. In *The Struggle for Ecological Democracy*, ed. Daniel Faber, 81–103. New York: Guilford.

Fitzgerald, Randall. 1992. The great spotted owl war. *Reader's Digest* 141(November): 91–95.

Foster, John Bellamy. 1993. The limits of environmentalism without class: Lessons from the ancient forest struggle of the Pacific Northwest. *Capitalism, Nature and Socialism* 4(1): 11–41.

Freudenberg, William, Lisa Wilson, and Daniel O'Leary. 1998. Forty years of spotted owls? A longitudinal analysis of logging industry job losses. *Sociological Perspectives* 41: 1–26.

Friedman, David, Jason Mark, Patricia Monahan, Carl Nash, and Clarance Ditlow. 2001. "Drilling in Detroit." Union of Concerned Scientists, June 2001. Cambridge, MA: UCS Publications.

Gais, Thomas, Mark Peterson, and Jack Walker. 1984. Interest groups, iron triangles and representative institutions in American national government. *British Journal of Political Science* 14: 161–185.

Gallup Poll Monthly. 1995. "Top Priority: Environment or Economy?" April 17–19. Survey GO 22-50001-017, Q. 23.

Gamson, William. 1961. A theory of coalition formation. *American Sociological Review* 26: 373–382.

Gamson, William. 1990. *The strategy of social protest*, 2nd ed. Belmont, CA: Wadsworth.

Gamson, William, and Bruce Fireman. 1979. Utilitarian logic in the resource mobilization perspective. In *The Dynamics of Social Movements*, ed. Mayer Zald and John McCarthy, 8–45. Cambridge, MA: Winthrop.

Gladwin, Thomas. 1980. Patterns of environmental conflict over industrial facilities in the United States, 1970–78. *Natural Resources Journal* 20: 243–274.

Goodstein, Eban. 1999. *The trade-off myth: Fact and fiction about jobs and the environment*. Washington, DC: Island.

Gordon, Robert. 1998. "Shell no!" OCAW and the labor-environmental alliance. *Environmental History* 3: 460–488.

Gottlieb, Robert. 1993. *Forcing the spring*. Washington, DC: Island.

Gottlieb, Robert. 2001. *Environmentalism unbound*. Cambridge: MIT Press.

Goudreau, Mike. 1991. Work and the Land. *Metroland* (Albany, NY), November 28–December 4.

Gouldner, Alvin. 1979. *The future of intellectuals and the rise of the new class*. New York: Seabury.

Gray, John S. 1995. Jobs and the environment. In *Let the people judge*, ed. John Echeverria and Raymond Booth Eby, 191–201. Washington, DC: Island.

Greenhouse, Steven. 1999. Longtime foes join to promote jobs and earth. *New York Times*, November 4, p. A12.

Greenhouse, Steven. 2000. More trouble for Teamsters president. *New York Times*, June 20, p. A12.

Greenhouse, Steven. 2001. Hearing Bush energy plan, union leaders offer praise. *New York Times*, May 15, p. A21.

Greider, William. 1992. *Who will tell the people?* New York: Simon & Schuster.

Grossman, Richard. 1985. Environmentalists and the labor movement. *Socialist Review* 82–88:63–87.

Grossman, Zoltan. 2001. "Let's not create evilness for this river": Interethnic environmental alliances of Native Americans and rural whites in northern Wisconsin." In *Forging radical alliances across difference*, ed. Jill Bystydzienski and Steven Schacht, 146–159. London: Rowman and Littlefield.

Gup, Ted. 1990. Owl vs man. *Time* 135(June 25): 56–62.

Hammond, Mark. 1990. Labor leaders, environmentalists discover they have similar goal. *The Daily Gazette* (Schenectady, NY), November 28.

Hathaway, Will, and David Meyer. 1997. Competition and cooperation in movement coalitions: Lobbying for peace in the 1980s. In *Coalitions and political movements*, ed. Thomas Rochon and David Meyer, 61–80. Boulder, CO: Rienner.

Head, Rebecca. 1995. Health based standards: What role environmental justice? In *Environmental Justice*, ed. Bunyan Bryant, 45–56. Washington, DC: Island.

Hinckley, Barbara. 1981. *Coalitions and politics.* New York: Harcourt Brace Jovanovich.

Hojnacki, Marie. 1997. Interest groups' decisions to join alliances of work alone. *American Journal of Political Science* 41(1): 61–87.

Huber, George. 1991. Organizational learning. *Organization Science* 2(1): 88–115.

Hula, Kevin W. 1999. *Lobbying together.* Washington, DC: Georgetown University Press.

Inglehart, Ronald. 1990. *Cultural shift in advanced industrial society.* Princeton, NJ: Princeton University Press.

Jenkins-Smith, Hank. 1991. Explaining change in policy subsystems: Analysis of coalition stability and defection over time. *American Journal of Political Science* 35: 851–880.

Jenkins-Smith, Hank, and Paul Sabatier. 1994. Evaluating the advocacy coalition framework. *Journal of Public Policy* 14: 175–203.

Johnston, Paul. 1994. *Success where others fail: Social movement unionism and the public workplace.* Ithaca, NY: ILR.

Jones, Robert Emmet, and Riley E. Dunlap. 1992. The social bases of environmental concern: Have they changed over time? *Rural Sociology* 57: 28–47.

Judge, Donald. 1995. The wise use threat to American workers. In *Let the People Judge*, ed. John Echeverria and Raymond Booth Eby, 202–207. Washington, DC: Island Press.

Kahn, Joseph. 2001. Core's split on energy is costly to democrats. *New York Times*, August 3, p. A20.

Kannar, George. 1993. Making the Teamsters safe for democracy. *Yale Law Review* 102: 1645–1687.

Kazis, Richard, and Richard Grossman. 1991. *Fear at work.* Philadelphia: New Society.

Keller, Bill. 1982. Coalitions and associations transform strategy, methods of lobbying in Washington. *Congressional Quarterly* 23 (January): 119–123.

Kelley, E. W. 1970. Utility theory and political coalitions." In *The Study of Coalition Behavior*, ed. Sven Groennings, E. W. Kelley, and Michael Leiserson, 466–481. New York: Holt, Rinehart and Winston.

Kelley, H. H., and A. J. Arrowood. 1960. Coalitions in the triad: Critique and experiment. *Sociometry* 23: 231–244.

Klandermans, Bert, and Dirk Oegema. 1987. Potentials, networks, motivations and barriers: Steps toward participation in social movements. *American Sociological Review* 52: 519–532.

Kleidman, Robert, and Thomas R. Rochon. 1997. Dilemmas in organization in peace campaigns. In *Coalitions and Political Movements: The Lessons of the Nuclear Freeze*, ed. Thomas R. Rochon and David S. Meyer, 47–60. Boulder: Rienner.

Larana, Enrique, Hank Johnston, and Joseph R. Gusfield. 1994. *New social movements: From ideology to identity*. Philadelphia: Temple University Press.

Larsen, Dana. 1995. Building broad-based coalitions to oppose takings legislation. In *Let the People Judge*, ed. John Echeverria and Raymond Booth Eby, 304–312. Washington, DC: Island.

Lawler, Edward, and George A. Young, Jr. 1975. Coalition formation: An integrative model. *Sociometry* 38: 1–17.

Leiserson, Michael. 1970. Game theory and the study of coalition behavior. In *The Study of Coalition Behavior*, ed. Sven Groennings, E. W. Kelley, and Michael Leiserson, 255–272. New York: Holt, Rinehart and Winston.

Lester, James, David Allen, and Kelly Hill. 2001. *Environmental injustice in the United States*. Boulder, CO: Westview.

Levitt, Barbara, and James March. 1988. Organizational learning. *Annual Review of Sociology* 14: 319–340.

Logan, Rebecca, and Dorothy Nelkin. 1980. Labor and nuclear power. *Environment* 22(2): 6–34.

Loomis, Burdett. 1986. Coalitions of interests: Building bridges in the Balkanized state. In *Interest group politics*, 2nd ed., ed. Allen Cigler and Burdett Loomis, 258–274. Washington, DC: CQ Press.

Lowi, Theodore. 1969. *The end of liberalism*. New York: Norton.

Lyons, Lenore. 2001. Negotiating difference: Singaporean women building an ethics of respect. In *Forging radical alliances across difference*, ed. Jill Bystydzienski and Steven Schacht, 177–190. London: Rowman and Littlefield.

MacArthur, John R. Jr. 2000. *The selling of free trade: NAFTA, Washington and the subversion of American democracy*. New York: Hill and Wang.

Madison, James. 1961. *The Federalist Papers*, no. 10, 77–84. New York: American Library.

Mann, Eric. 1990. Labor-community coalitions as a tactic for labor insurgency. In *Building bridges: the emerging grassroots coalition of labor and community*, ed. Jeremy Brecher and Tim Costello, 113–132. New York: Monthly Review.

March, James, and Herbert Simon. 1958. *Organizations.* New York: John Wiley.

Marwell, Gerald, and D. Schmitt. 1975. *Cooperation.* New York: Academic.

Marx, Karl. 1967. *Capital,* vol. 1. New York: International.

Mathis, Benton J., Leigh C. Lawson, and Christopher Parker. 1993. Labor law. *Mercer Law Review* 44: 1251–1279.

Mazza, Dave. 1993. *God, land and politics: The wise use and Christian right connection in 1992 Oregon politics.* Portland, OR: Wise Use Public Exposure Project.

McAdam, Doug. 1983. Tactical innovation and the pace of insurgency. *American Sociological Review* 48(6): 735–754.

McCarthy, John D., David W. Britt, and Mark Wolfson. 1991. The institutional channeling of social movements by the state in the United States. *Research in Social Movements, Conflicts and Change* 13: 45–76.

McCarthy, John D., and Mayer N. Zald. 1973. *The trend of social movements in America: Professionalization and resource mobilization.* Morristown, NJ: General Learning.

McCarthy, John D., and Mayer N. Zald. 1977. Resource mobilization and social movements: A partial theory. *American Journal of Sociology* 82: 1212–1293.

McClure, Laura. 1992. Meet the labor movement. *Environmental Action* 24(1): 14–23.

Meissner, Werner. 1986. Employment, income and welfare implications of environmental policy. In *Distributional conflicts in environmental-resource policy,* ed. Allan Schnaiberg, Nicholas Watts, and Klaus Zimmerman, 38–48. Berlin: Gower.

Melucci, Alberto. 1980. The new social movements: A theoretical approach. *Social Science Information* 19: 199–226.

Melucci, Alberto. 1994. A strange kind of newness: What's "New" in new social movements? In *New Social Movements: From Ideology to Identity,* ed. Enrique Larana, Hank Johnston and Joseph R. Gusfield, 101–130. Philadelphia: Temple University Press.

Meyer, Stephen. 1992. Environmentalism and economic prosperity: Testing the environmental impact Hypothesis. Massachusetts Institute of Technology Project on Environmental Politics and Policy, Cambridge, Massachusetts.

Meyer, Stephen. 1993. "Environmentalism and economic prosperity: An update." Massachusetts Institute of Technology Project on Environmental Politics and Policy, Cambridge, Massachusetts.

Michels, Robert. 1949. *Political parties.* Glencoe, IL: Free Press.

Miliband, Ralph. 1983. State power and class interests. *New Left Review* 138(March–April): 57–68.

Miller, Alan. 1980. Towards an environmental/labor coalition. *Environment* 22(5): 32–39.

Mills, C. Wright. 1956. *The power elite.* London: Oxford University Press.

Mills, Mike. 1990. *CQ Weekly* 48(43): 3589.

Moberg, David. 1999. Greens and labor: It's a coalition that gives corporate polluters fits. *Sierra* (January/February): 46–52.

Moberg, David. 2000. For unions, Green's not easy. *The Nation* 270(7): 17–22.

Moberg, David. 2002. Enough blue-green bickering. *In These Times*, May 7.

Moe, Terry M. 1980. *The organization of interests*. Chicago: University of Chicago Press.

Morrison, Denton. 1986. How and why environmental consciousness has trickled down. In *Distributional Conflicts in Environmental-Resource Policy*, ed. Allan Schnaiberg, Nicholas Watts, and Klaus Zimmerman, 187–220. Berlin: Gower.

Morrison, Denton, Kenneth E. Hornback, and W. Keith Warner. 1972. The environmental movement: Some preliminary observations and prediction. In *Social Behavior, Natural Resources, and the Environment*, ed. W. R. Rauch Jr., N. H. Cheek Jr., and L. Taylor, 259–279. New York: Harper and Row.

Morriss, Andrew. 2000. The politics of the Clean Air Act. In *Political environmentalism*, ed. Terry Anderson, 263–318. Stanford, CA: Hoover Institution Press.

Muller, Edward, and Karl Dieter Opp. 1986. Rational choice and rebellious collective action. *American Political Science Review* 80: 471–488.

Muller, Joann, with Kathleen Kerwin and David Welch. 2002. A new industry. *Business Week* (July 15): 98.

Murphy, Raymond. 1994. *Rationality and nature*. Boulder, CO: Westview.

Nissen, Bruce. 1995. *Fighting for jobs and the environment*. Albany: State University of New York Press.

Obach, Brian K. 1994. Social change orientation: Personal vs. institutional conceptions of social change and their relation to tactics and group participation. Master's thesis, University of Wisconsin, Madison.

Obach, Brian K. 1999. The Wisconsin labor-environmental network: A case study of coalition formation among organized labor and the environmental movement. *Organization and Environment* 12(1): 45–74.

Obach, Brian K. 2002. Labor-environmental relations: An analysis of the relationship between labor unions and environmentalists. *Social Science Quarterly* 83(1): 82–100.

Offe, Claus. 1987. Challenging the boundaries of institutional politics: Social movements since the 1960s. In *Changing Boundaries of the Political*, ed. C. Maire, 63–105. Cambridge: Cambridge University Press.

Oil, Chemical, and Atomic Workers International Union. n.d. *Understanding the conflict between jobs and the environment*. Denver: Author.

Oliver, Pamela E., and Gerald Marwell. 1992. Mobilizing technologies for collective action. In *Frontiers in Social Movement Theory*, ed. Aldon Morris and Carol McClurg Mueller, 251–272. New Haven: Yale University Press.

Olson, Mancur. 1965. *The logic of collective action: Public goods and the theory of groups.* Cambridge: Harvard University Press.

Opp, Karl-Dieter. 1990. Postmaterialism, collective action, and political protest. *American Journal of Political Science* 34: 212–235.

Organisation for Economic Cooperation and Development (OECD). 1997. *Environmental Policies and Employment.* Paris: Author.

Piven, Frances Fox, and Richard Cloward. 1977. *Poor people's movements.* New York: Pantheon.

Polsby, Nelson. 1978. Coalition and faction in American politics: An institutional view. In *Emerging coalitions in American politics,* ed. Seymore Martin Lipset, 103–126. San Francisco: Institute for Contemporary Studies.

Potier, Michael. 1986. Capital and labour reallocation in the face of environmental policy. In *Distributional conflicts in environmental-resource policy,* ed. Allan Schnaiberg, Nicholas Watts, and Klaus Zimmerman, 253–271. Berlin: Gower.

Poulantzas, Nicos. 1969. The problem of the Capitalist state. *New Left Review* 58: 67–78.

Przeworski, Adam. 1985. *Capitalism and social democracy.* Cambridge: Cambridge University Press.

Pulp and Paperworkers Resource Council (PPRC). n.d. "History of the PPRC." Available on the PPRC web page at <www. PPRC.com>.

Putnam, Robert. 2000. *Bowling alone.* New York: Simon and Schuster.

Ramos, Tarso. 1995. Wise use in the West: The case of the Northwest timber industry. In *Let the people judge,* ed. John Echeverria and Raymond Booth Eby, 82–118. Washington, DC: Island.

Richards, Barbara. 1990. The community-labor alliance of New Haven. In *Building bridges: The emerging grassroots coalition of labor and community,* ed. Jeremy Brecher and Tim Costello, 70–89. New York: Monthly Review.

Riker, William. 1962. *The theory of political coalitions.* New Haven: Yale University Press.

Ringquist, Evan J. 1995. Environmental protection regulation. In *Regulation and consumer protection,* 2nd ed., ed. Kenneth J. Meier and E. Thomas Garman, 147–196. Houston: Dame.

Ripley, Randall, and Grace Franklin. 1984. *Congress, the bureaucracy and public policy.* Homewood, IL: Dorsey.

Rose, Fred. 1997. Toward a class-cultural theory of social movements: Reinterpreting new social movements. *Sociological Forum* 12: 461–494.

Rose, Fred. 2000. *Coalitions across the class divide.* Ithaca:Cornell University Press.

Ruben, Barbara. 1992. Root rot. *Environmental Action* 24: 25–30.

Rucht, Dieter. 1999. Linking organization and mobilization: Michel's iron law of oligarchy reconsidered. *Mobilization: An International Journal* 4(2): 151–169.

Sabatier, Paul. 1988. An advocacy coalition framework of policy change and the role of policy oriented learning therein. *Policy Sciences* 21: 129–168.

Sabatier, Paul, John Loomis, and Catherine McCarthy. 1995. Hierarchical controls, professional norms, local constituencies and budget maximization: An Analysis of US Forest Service planning decisions. *American Journal of Political Science* 39: 204–242.

Sabatier, Paul, and Susan McLaughlin. 1990. Belief congruence between interest group leaders and members: An empirical analysis of three theories and a suggested synthesis. *Journal of Politics* 52: 914–935.

Salisbury, Robert. 1969. An exchange theory of interest groups. *Midwest Journal of Political Science* 13: 1–32.

Salisbury, Robert. 1990. The paradox of interest groups in Washington: More groups, less clout. In *The new American political system*, ed. Antony King, 203–230. Washington, DC: AEI Press.

Salisbury, Robert, John Heinz, Edward Laumann, and Robert Nelson. 1987. Who works with whom? Interest group alliances and opposition. *American Political Science Review* 81: 1217–1234.

Sanchez, Samantha. 1996. How the West is won. *American Prospect* (March/April): 37–42.

Schaeffer, Robert. 1997. *Understanding globalization*. New York: Rowman and Littlefield.

Schattschneider, E. E. 1960. *The semi-sovereign people*. New York: Holt, Rinehart and Winston.

Schlozman, Kay Lehman, and John Tierney. 1986. *Organized interests and American democracy*. New York: Harper and Row.

Schnaiberg, Allan. 1980. *The environment: From surplus to scarcity*. New York: Oxford University Press.

Scott, W. Richard. 1992. *Organizations: Rational, natural, and open systems*. Englewood Cliffs, NJ: Prentice-Hall.

Seidman, Gay. 1994. *Manufacturing militance*. Berkeley and Los Angeles: University of California Press.

Siegmann, Heinrich. 1985. *The conflict between labor and environmentalism in the Federal Republic of Germany and the United States*. New York: St. Martin's.

Siegmann, Heinrich. 1986. Environmental policy and trade unions in the United States. In *Distributional conflicts in environmental-resource policy*, ed. Allan Schnaiberg, Nicholas Watts, and Klaus Zimmerman, 315–327. Berlin: Gower.

Sierra Club. 2002. Increasing America's fuel economy, February. ⟨www.sierraclub.org/globalwarming⟩.

Simmons, Louise. 1990. Organizational and leadership models in labor-community coalitions. In *Building bridges: The emerging grassroots coalition of labor and community*, ed. Jeremy Brecher and Tim Costello, 229–233. New York: Monthly Review.

Simon, Herbert. 1976. *Administrative behavior*. 3rd ed. New York: Macmillan.

Sink, David. 1991. Transorganizational development in urban policy coalitions. *Human Relations* 44: 1179–1195.

Smith, Liv. 1980. Labor and the no-nukes movement. *Socialist Review* no. 54 (November–December): 135–149.

Snow, David, and Robert Benford. 1988. Ideology, frame resonance, and participant mobilization. In *From structure to action: Comparing social movement research across cultures*, vol. 1 in *International social movement research*, ed. B. Klandermans, H. Kriese, and S. Tarrow, 197–218. Greenwich, CT: JAI.

Snow, David, and Robert Benford. 1992. Master frames and cycles of protest. In *Frontiers in social movement theory*, ed. Aldon Morris and Carol Mueller, 133–155. New Haven: Yale University Press.

Snow, David, E. Burke Rocheford, Jr., Steven K. Worden, and Robert Benford. 1986. Frame alignment processes, micromobilization and movement participation. *American Sociological Review* 51(4): 464–481.

Staggenborg, Suzanne. 1986. Coalition work in the pro-choice movement. *Social Problems* 33(5): 374–389.

Staggenborg, Suzanne. 1988. The consequences of professionalization and formalization in the pro-choice movement. *American Sociological Review* 53: 585–606.

Starr, Jerold M. 2001. The challenge and rewards of coalition building: Pittsburgh's Alliance for Progressive Action. In *Forging radical alliances across difference*, ed. Jill Bystydzienski and Steven Schacht, 107–119. London: Rowman and Littlefield.

Steedly, Homer, and John Foley. 1990. The success of protest groups: Multivariate analyses. Appendix A in *The Strategy of Social Protest*, ed. William Gamson, 182–198. Belmont, CA: Wadsworth.

Stinchcombe, Arthur. 1965. Social structure and organizations. In *Handbook of organizations*, ed. James March, 142–193. Chicago: Rand McNally.

Summers, Clyde. 1977. The individual employee's rights under the collective bargaining agreement: What constitutes fair representation? *University of Pennsylvania Law Review* 126.

Taylor, Verta, and Nancy Whittier. 1992. Collective identity in social movement communities: Lesbian feminist mobilization. In *Frontiers in social movement theory*, ed. Aldon Morris and Carol McClurg Mueller, 104–130. New Haven: Yale University Press.

Teamsters and turtles in [temporary?] tiff. *The American Prospect* 13(7): 8.

Thomas, Clive, and Ronald Hrebenar. 1994. Interest group power in state politics: A complex phenomenon. *Comparative State Politics* 15(3): 7–18.

Tillman, Ray M., and Michael S. Cummings. 1999. *The transformation of US unions*. Boulder, CO: Rienner.

Tilly, Charles. 1978. *From mobilization to revolution*. Englewood Cliffs, NJ: Prentice Hall.

Tocqueville, Alexis de. 1990. *Democracy in America*. New York: Vintage.

Truman, David. 1951. *The governmental process*. New York: Knopf.

United Auto Workers (UAW) Community Action Program. 2002. Energy issues CAPE. <www. uaw.org/cap/oz/issue/issue07.html> (position paper on Corporate Average Fuel Economy).

United Church of Christ Commission for Racial Justice. 1987. *Toxic wastes and race in the United States: A national study of the racial and socioeconomic characteristics of communities with hazardous waste sites*. New York: United Church of Christ.

United States Bureau of the Census. 1997. *Statistical abstracts of the United States*. Washington, DC: United States Government Printing Office.

United States Bureau of the Census. 1998. *County business patterns 1996*. Washington, DC: United States Government Printing Office.

United States Environmental Protection Agency. 1999. The benefits and costs of the Clean Air Act Amendments of 1990. (Report to Congress updated, no. EPA-410-R-99001).

Van, Liere, and Riley Dunlap. 1980. The social bases of environmental concern: A review of hypotheses, explanations and empirical evidence. *Public Opinion Quarterly* 44(2): 181–197.

Verba, Sidney, and Norman Nie. 1972. *Participation in America: Political democracy and social equality*. New York: Harper and Row.

Von Neumann, John, and Oskar Morgenstern. 1947. *Theory of games and economic behavior*. 2nd ed. Princeton, NJ: Princeton University Press.

Walker, Jack L. 1983. The origins and maintenance of interest groups in America. *American Political Science Review* 77: 390–406.

Wattenberg, Martin. 1984. *The decline of American political parties, 1952–1980*. Cambridge: Harvard University Press.

Watts, Nicholas. 1986. From consensus to dissensus: The role of distributional conflicts in environmental-resource policy. In *Distributional conflicts in environmental-resource policy*, ed. Allan Schnaiberg, Nicholas Watts, and Klaus Zimmerman, 1–14. Berlin: Gower.

Westra, Laura, and Peter Wenz. 1995. *Faces of environmental racism*. Lanham, MD: Rowman and Littlefield.

Willis, Paul. 1981. *Learning to labor: How working class kids get working class jobs*. New York: Teachers College Press.

Wilson, Graham. 1990. *Interest groups*. Cambridge, MA: Basil Blackwell.

Wilson, Graham. 1993. American interest groups. In *Pressure groups*, ed. Jeremy Richardson, 131–145. New York: Oxford University Press.

Wilson, James Q. 1973. *Political organizations*. New York: Basic.

Wright, Erik Olin. 1976. Class boundaries in advanced capitalist societies. *New Left Review* 98: 3–41.

Wright, John R., and Arthur Goldberg. 1985. Risk and uncertainty as factors in durability of political coalitions. *American Political Science Review* 79: 704–718.

Yandle, Bruce. 1983. Bootleggers and Baptists: The education of a regulatory economist. *Regulation* 7(3): 12.

Yandle, Bruce. 1985. Unions and environmental regulation. *Journal of Labor Research* 6: 429–436.

Yin, Robert. 1994. *Case study research: Design and methods*. 2nd ed. Thousand Oaks, CA: Sage.

Zald, Mayer, and Roberta Ash. 1966. Social movement organizations: Growth, decay and change. *Social Forces* 44: 327–341.

Zald, Mayer, and John D. McCarthy. 1980. Social movement industries: Competition and cooperation among movement organizations. *Research in Social Movements, Conflicts and Change* 3: 1–20.

Zimmerman, Klaus. 1986. Distributional considerations and the environmental policy process. In *Distributional conflicts in environmental-resource policy*, ed. Allan Schnaiberg, Nicholas Watts, and Klaus Zimmerman, 95–108. Berlin: Gower.

Index